AF696046

Modernising Agrifood Chains in China

Modernising Agrifood Chains in China: Implications for Rural Development

By

Scott Waldron

CAMBRIDGE SCHOLARS PUBLISHING

Modernising Agrifood Chains in China:
Implications for Rural Development,
by Scott Waldron

This book first published 2010

Cambridge Scholars Publishing

12 Back Chapman Street, Newcastle upon Tyne, NE6 2XX, UK

British Library Cataloguing in Publication Data
A catalogue record for this book is available from the British Library

Copyright © 2010 by Scott Waldron

All rights for this book reserved. No part of this book may be reproduced, stored in a retrieval system, or transmitted, in any form or by any means, electronic, mechanical, photocopying, recording or otherwise, without the prior permission of the copyright owner.

ISBN (10): 1-4438-2507-7, ISBN (13): 978-1-4438-2507-8

Dedicated to Mum and Dad
For the farm, formal and life education
that made this research possible

And in loving memory of
Olive

Contents

List of figures

List of boxes

List of Tables

Foreword

China faces enormous challenges in generating sustained and inclusive agricultural and rural development. But rapid growth occurring both inside and outside the agricultural sector provides new opportunities to meet the challenges. The development of higher value agrifood chains may provide opportunities for rural households to increase their levels of specialisation, scale and incomes, and for rural areas to broaden their employment and tax bases.

Facilitation of these developments effectively forms an agricultural modernisation strategy that is becoming increasingly described and prescribed by Chinese policy makers, researchers and development agencies. Despite the attention that has been given to this agricultural modernisation strategy, it has not proceeded without problems and has been subject to little rigorous scrutiny. This study uses a detailed industry case study approach to examine the way that Chinese agriculture is modernising and, as an applied study, to identify areas where China's agricultural modernisation strategy can be refined.

China's agricultural modernisation strategy is analysed through the window of a particular Chinese agricultural and rural industry, namely the beef industry. The beef industry resembles many other agricultural and livestock industries that are commercialising, modernising and segmenting with important implications for rural incomes, food safety and the environment. In addition to providing insights into developments in the broader agricultural and livestock sectors, the Chinese beef industry is also worth examining in its own right because of its significant place in the world beef industry and in China's livestock sector, and because it is used as a "pillar industry" in the development of many poor areas. While the study focuses predominantly on the beef industry, differences and similarities with other livestock industries are noted throughout the analysis.

The use of an industry case study allows for a detailed analysis of a cross section of China's retail, processing, marketing and production sectors, and the integration of these sectors that form the industry. One novel aspect of the study is that the industry modernisation process is seen as a movement from low to high value agrifood chains in all of these sectors and for the industry as a whole. The study examines the drivers of, and constraints to, the movement from low to high value agrifood chains, the sustainability of the development of the higher value agrifood chains, and the implications for industry participants.

The study draws on multiple forms of data, including extensive interview and fieldwork data over a ten year period across China, budget and scenario analyses; and macro-level data and policy documents. The cross-verification of this data allows for a rigorous, multi-disciplinary and grounded form of analysis that can complement other studies that draw on fewer sources of data. The study also provides a longitudinal and diachronic perspective of the agricultural modernisation process in contrast to a shorter term snapshot.

The study examines the reasons and the measures by which China is attempting to move the agricultural sector beyond its primary reliance on traditional, low value agrifood chains. While the logic of these reasons and measures can be discerned, it is argued that China has attempted to "skip" development stages by over-emphasising the development of large scale modern high value agrifood chains. The attempt to force the pace of modernisation has led to significant distortions and detrimental impacts. The study argues the need for China to refine its agricultural modernisation development strategy to be both more market conforming and to bring more benefits to industry participants in rural areas. This is best done by taking a more incremental approach to agricultural modernisation with an increased emphasis on mid value agrifood chains and the development of market facilitating policies. The analysis, findings and recommendations of the study may be of interest to policy makers, development agencies and researchers working on China's agricultural and livestock sectors.

Acknowledgements

This book is based on a collaborative research program conducted over many years so an enormous amount of debt and gratitude lies behind it.

The research draws heavily on the methods, knowledge, contacts and mentoring of Colin Brown and John Longworth. I have been fortunate to be able to observe on a day to day basis the way that Colin and John deal with problems and conduct "muddy boots" agricultural economics research, and it has been a privilege to work, travel and share a wide range of experiences with colleagues of the intelligence and integrity of Colin and John.

While I have a background in Chinese studies, much of what I have learnt about China has been through Chinese collaborators who have become great friends including Zhao Yutian, Liu Yuman, Zhang Cungen, Lu Xiaoping, Li Suoping, Lin Xiangjin, Li Ping, Li Wei, Chai Jun and Jiang Hongmao. Plenty of ideas, stories and travel have also been shared with Ben Lyons, Ralph van Gelder, Ian Jarratt, Dominic Smith, Cai Haiou and Stephanie Cash. While not directly involved in the research for this book, it draws on the knowledge and interest in rural China of Greg Veeck, Niu Ruofeng, Ke Bingsheng, Scott Rozelle, Allan Rae, Masaru Kagatsume, David Schak and Colin Mackerras.

The work was made possible through the support of the Australian Centre of International Agricultural Research (in which Ken Menz was instrumental) and Meat and Livestock Australia (thanks especially to Tim Kelf). These agencies and people provided ongoing support for this beef research and also for subsequent areas of research together with Australian Wool Innovations and the (Australian) Department of Agriculture, Forestry and Fisheries.

Thanks to the Australian Agricultural and Resource Economics Society for awarding the thesis on which this book is based the Best PhD Thesis Prize of 2009, to Cambridge Scholars Publishing for publishing the book with the thesis content intact, and to Sunset Digital for formatting the manuscript to book standard.

The School of Natural and Rural Systems Management at The University of Queensland has provided excellent research services and facilities. Long term staff members of the School especially Rob Cramb and Mal Wegener with many years of experience in agriculture and rural development have built a collegiate environment that is a real pleasure to work in.

One of the privileges of the research and work has been the opportunity to conduct fieldwork over long periods and in many parts of rural China. Special thanks are due to the Ministry of Agriculture and Animal Husbandry Bureaus from central down to county levels for hosting the visits and extending hospitality well beyond the call of duty. Increasingly busy government officials, business people, researchers and households have given up time, information, knowledge and other resources for the research. There are too many areas and people to thank individually, but hopefully some of the ideas in this book can contribute to the debate about how the beef and other rural industries might develop into the future.

Family is important for everything and no less so in seeing through tasks like writing theses and books. Lygia and Zoe were particularly understanding, especially as Leon was delivered at the same time as the thesis, when Maria was also a great help.

The book is dedicated to my parents Jenny and Tony and in loving memory of my grandmother Olive.

Chapter One

Introduction

Rapid economic development has delivered new opportunities and challenges for agricultural and rural development in China. Meeting these opportunities and challenges requires new and increasingly refined strategies that must be underpinned by detailed and pragmatic forms of analysis. This book aims to provide such a form of analysis through a detailed rural industry case study, which may be of interest to policy makers, development agencies and researchers working on China's agricultural and livestock sectors.

1.1 Background and rationale

China is undergoing a transformation of historic proportions. An annual GDP growth rate of nearly 10% over the thirty years of the reform era (National Bureau of Statistics, various years) has bought about the largest reduction in poverty in history (World Bank, 2005), the largest movement of humans in peacetime history (from rural to urban areas) and an accelerated transition from an agrarian to an industrial nation (Rozelle et al., 2002).

Despite the enormous progress, however, income gaps in China are now amongst the most unequal in the world (Chang, 2002). The fault lines of inequality lie between rural and urban people and rural and urban areas. The lagging of the agricultural sector, rural incomes and rural areas—known in China as the *sannong* or "three rurals problem"—is regularly cited as the biggest development challenge facing China today.[1]

Much of the progress in solving the latter two of the "rurals problem"—rural incomes and rural area development—has not been achieved through the first of the "rurals problem"—of agricultural development—but through

[1] The term is *sannong* ("the three rurals") is an abbreviation for the three terms *nongye, nongmin* and *nongcun*, which share the root character *nong* (rural). Interpreted from the perspective of this study, the term *nongye* refers to agriculture; *nongmin* refers to rural people or households; and *nongcun* refers to rural areas. The "three rurals" issue has been cited as the major development challenge in China at nearly all major occasions in recent years: from Zhu Rongji's address to the National People's Congress in March 2002; to the Number One Document of 2004; to the 11th Five Year Plan (2006–2010).

the development of rural industry and the absorption of rural labour into the urban, industry and service sectors. Almost half of the increase in income of farmers in China between 2000 and 2002 came from jobs outside their own townships (Ministry of Agriculture, 2005). To facilitate this so called *zhuanyi* process, China has introduced measures to increase resource mobility for labour (work migration and household registration) and land (the on-leasing of agricultural land). Chinese policy makers believe that further reforms in this direction will "convert" rural producers into urban consumers, while a less crowded agricultural landscape will provide "room" for those rural households that remain in agriculture to increase returns from agricultural activities.

While continuation of this *zhuanyi* process may contribute to solving the "three rurals problem" over the longer term, China continues to take a highly proactive approach to agricultural development. This is not just because of China's preoccupation with food security and mercantilist trade objectives. Although the share of agriculture in China's economy decreased from 30% in 1985 to 13% in 2005, 62% of the value from the light industry sector was derived from products that use agricultural inputs (Ministry of Agriculture, 2004). Similarly, while the proportion of the Chinese working population engaged in agriculture decreased from 62% in 1985 to 45% in 2005, another 25% of China's working population work in the 12 sectors that use agricultural products as inputs (Ministry of Agriculture, 2006; Township and Village Enterprise Bureau, 2007, personal communication)[2]. Far from ignoring the agricultural sector, policy makers continue to see the "three rurals problem" as interconnected, where agricultural development remains central to bringing about rural and regional development.

In this regard, Chinese policy makers are optimistic about and actively promote several trends in Chinese agriculture including: a) the growing market for higher value agricultural products due to income growth, urbanisation and increased access to export markets; b) the modernisation of the agribusiness sector, especially processing and marketing; and c) that these trends in the off-farm sector will bring opportunities for rural households to sell higher priced farm outputs and to increase farm specialisation and scale. It is also envisioned that the trends will broaden the employment and tax revenue bases of rural areas. A simple representation of these trends appears in Figure 1.1.

[2] The sectors include food products, food processing, beverages, tobacco, textiles, clothing and footwear, leather processing, timber, furniture, paper products and printing. Employment in activities based around agricultural products would be higher if services (such as food service, trading, retailing and transport) were included.

The facilitation of these trends effectively forms an agricultural and rural development strategy that is becoming increasingly described and prescribed by researchers and development agencies (see Chapter 2) and by Chinese policy makers (see Chapter 4). Researchers and development agencies have turned attention to a series of so called "revolutions" including the "high value agriculture revolution", the "agribusiness revolution", the "small-holder productivity revolution" and the "livestock revolution". Similarly, Chinese policy makers at all levels strongly support the development of agro-industrial "dragon head" enterprises because of their perceived capacity to "lead" (*daidong*) the "scattered" (*fensan*) rural households and rural groups through the "length of the dragon" (the supply chain) into higher value and distant end markets. Economic agents and consumers are driving other major trends in Chinese agriculture including market segmentation, product differentiation, specialisation and supply chain integration. The facilitation of these trends is referred to in this book as an "agricultural modernisation strategy", in line with the "Modern Agriculture" theme of China's Number One Document of 2007 (jointly issued by the Chinese Communist Party Central Committee and the State Council) and in the Third Plenary Session of the 17th Chinese Communist Party Central Committee of September 2008.

Despite the attention that has been given to this agricultural modernisation strategy, it has certainly not proceeded without problems in China and has been subject to surprisingly little detailed, critical scrutiny either in the literature or by Chinese policy makers. This study aims to show how the modernisation strategy is progressing in practice and, as an applied study, to propose measures that may help refine it.

China's agricultural modernisation strategy is analysed through a focus on the Chinese livestock sector and on the beef industry in particular. The livestock sector and beef industry provides a useful case study for several reasons. Attention has turned in recent years to the growth in demand for animal protein and other animal products especially in developing countries. Rapid growth of the Chinese livestock sector has placed it at the forefront of the "international livestock revolution" with wide ranging development implications (Delgado *et al.*, 1999b). Of the 210 million rural households in China, 100 million households raise livestock, which account for up to 30% of farmers' incomes (China.org.cn, 2006). Within the livestock sector, the beef industry also continues to commercialise and impact on the livelihoods of more than 13 million cattle raising households and approximately 500,000 people in the agribusiness activities of trading and processing. As a result, the beef industry is used by policy makers in many poor areas of China as an instrument to pursue agricultural, rural and regional development.

Thus, the beef industry is worthwhile studying not only to provide insights into China's broader agricultural development and modernisation process, but also in its own right in terms of its potential contribution to livelihoods.

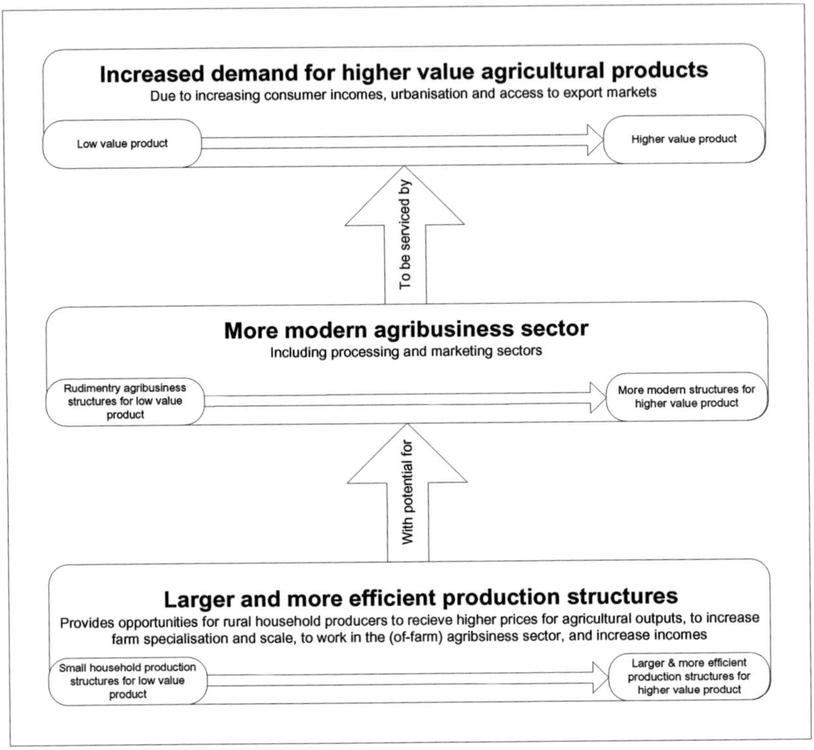

Figure 1.1: A simplified representation of an agricultural and rural development strategy in China

1.2 Research objectives

This study draws on a detailed industry analysis to examine the way that Chinese agriculture is modernising and developing, and to identify areas where China's agricultural modernisation strategy can be refined. Within this broad objective, three specific research questions are addressed:

1. What are the major trends occurring in the Chinese agricultural, livestock and beef sectors? *(This question is addressed through an analysis of*

macro policies and statistics in Chapters 4 and 5 of the study and also acts as context for more detailed analysis in the remainder of the study.)

2. How are the beef consumption, processing, marketing and production sectors developing and modernising? *(This question is addressed especially by analysing the incentives of economic agents and the policy environment in which they act in Chapters 6 to 9.)*
3. What policy measures should be taken to align China's modernisation strategy more closely with market developments and the interests of industry participants in rural areas? *(This question is addressed in Chapter 10 with a focus on different value chains, especially for the beef industry but also with relevance to the broader livestock and agricultural industries.)*

1.3 Methodology

China's broad agricultural modernisation strategy outlined above is investigated through a detailed industry case study approach. The industry is decomposed into a series of sectors and value chains which together form an "industry system" comprised of a series of inter-related sub-systems. The sub-systems and the relationships between them are examined through a diverse set of analytical tools, thus forming a multi-disciplinary and integrated study. The section below outlines briefly the methodology to be used in the study, which is expanded on in Chapter 3.

An industry approach

China's agricultural modernisation strategy is analysed through the window of a particular Chinese agricultural and rural industry, namely the beef industry. Using an industry as a unit of analysis has a number of benefits. First, the industry traverses the full agricultural and rural landscape comprised of markets, processing, marketing and production sectors, extension and services, and policies and institutions. An industry analysis therefore allows for an examination of a cross section of this landscape and the relationships in it, while at the same time limiting the scope of analysis to a manageable level. Second, an industry level analysis provides a window from which to examine broader economic phenomenon, including economic agents and their relationships, commodity flows and transactions, returns and rents extracted along the supply chain, organisation and technologies.

Third, the industry focus acts as a meso-level of analysis that links the macro-level (policy, institutional and the broader economic environment) with the micro-level (households and other economic agents). This can aid in the development of causal explanations of how macro settings impact at the micro-level. Fourth, industry level analysis allows for a detailed examination of the mechanics of economic activity, thus revealing issues and nuances that can be overlooked in more macro-level studies. Fifth, and most importantly in the context of this study, Chinese policy makers and international agencies very often use particular agricultural industries, industry targeting and industry specific policies as tools to pursue rural development objectives. The industry study therefore allows for examination of these industry development strategies and policies.

An industry case study approach also involves potential drawbacks, especially in generalising the approach and findings. There are considerable differences between agricultural industries and indeed between livestock industries in China. Different industries have different supply chains and structures and are impacted differently by cross-sectoral policies. These differences preclude "blanket" findings and recommendations across the agricultural sector as a whole (conversely this is also precisely the reason why industry studies are useful). At the same time, however, many Chinese agricultural industries are undergoing the type of modernisation processes seen in the beef industry with similar implications for industry participants. In order to generalise the approach and findings, discussion throughout the study places beef in context with the broader livestock sector and, where possible, makes explicit the differences between the beef and other livestock industries in which the author has comparable knowledge.

A systems approach and study framework

The objectives of this study are examined in the context of major trends in Chinese agriculture such as market segmentation, product differentiation, specialisation and increasing links between the agribusiness and on-farm sectors. This requires that the components of the industry are both "disassembled" and ordered. This study does so by treating the industry as a system comprised of a series of inter-related sub-systems. The "industry system" is encapsulated in the study framework presented in Figure 1.2, which also acts as a chapter outline of the study.

The left hand column of Figure 1.2 shows the chapter structure of the study. This initial introductory chapter (Chapter 1) is followed by the literature review (Chapter 2) and methodology chapter (Chapter 3). The subsequent two chapters draw on secondary information to place beef in context with the agricultural sector (Chapter 4) and to provide an overview of the beef industry (Chapter 5) that serves as context for the remaining chapters. Chapters 6 to 9 are based on primary information and more detailed analysis. Readers particularly interested in detailed analysis are encouraged to read the many footnotes of the book, which have been excluded from the main text in order to explicate arguments.

A central construct used in Chapters 6 to 9 is the dissection of the industry into various agrifood value chains. This segmentation dimension is designed to capture increasingly important aspects of Chinese agriculture where increasingly demanding consumers (end users) and customers (such as processors, retailers or traders) require increasingly differentiated agricultural products, which impacts on the structures of agrifood chains and their participants. There are major implications for agricultural, rural and regional development.

The study distinguishes between low, mid and high value agrifood chains, which are organised horizontally across Figure 1.2. These agrifood chains are illustrated as discrete agrifood chains in Figure 1.2 flowing vertically and linked by black arrows. However, in reality they form a continuum with significant product flows and therefore relationships between the chains. These are indicated by the horizontal and oblique shaded arrows between the boxes in Figure 3.1, which is a more elaborate depiction of the study framework.

These three agrifood chains flow through industry sectors which are listed vertically down Figure 1.2. The sectors cover both off-farm and on-farm sectors of the industry including retail (Chapter 6), processing (Chapter 7), marketing (Chapter 8) and production (Chapter 9). The chapters and sectors are ordered from downstream to upstream components of the industry on the basis that demand-pull side factors have a major impact on other aspects of the industry, although the impact on supply-push factors are also discussed.

The horizontal and vertical dimensions of the framework combine to form a matrix or the "industry system". At each node of the matrix lie specific industry sub-systems, including categories of industry participants (such as unspecialised cattle producers) or industry structures (such as mid value beef markets). Each of these constructs are analysed in specific sections of each chapter using analytical tools overviewed below. The study also examines the relationships between the sub-systems to show how issues or problems in one sub-system affect other sub-systems (for example how problems in modern processing affect small-holders). This is particularly

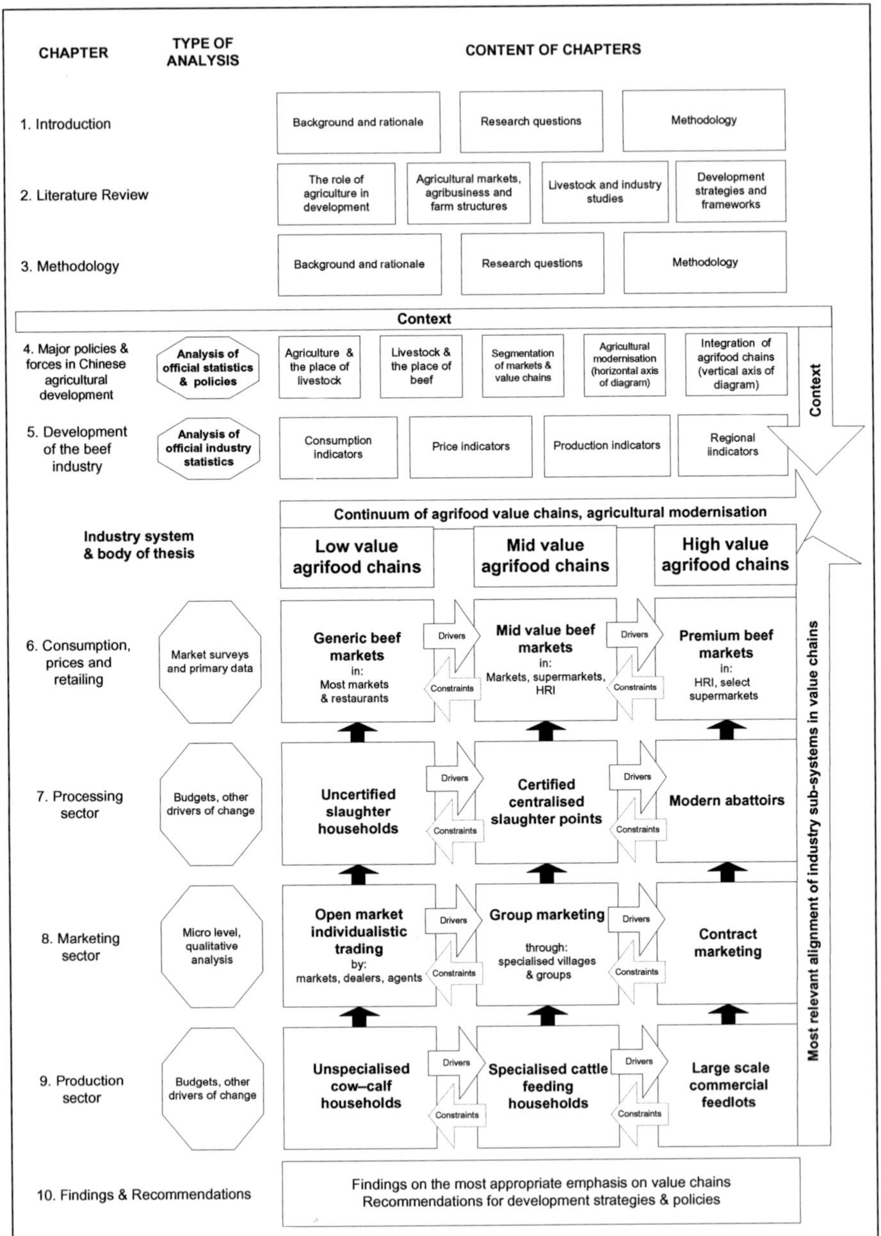

Figure 1.2: Outline of the beef industry system and links with book structure

important in discussing the implications of industry development for industry participants and in providing recommendations on how to address problems.

Another dimension of the framework is the relationships and movement between sub-systems on a horizontal level within each industry sector. The arrows pointing right represent drivers of industry modernisation, while the dashed arrows pointing left represent constraints. The drivers and the constraints can be both policy or economic related and are discussed in detail in Chapters 6 to 9. This analysis adds a transitionary or diachronic dimension to the framework that is important in examining the modernisation process and the implications for industry participants, and helps frame development options for industry participants and policy makers that are discussed in the concluding chapter of the study.

A multi-disciplinary approach

The "industry system" as depicted by the framework imposes the analytical boundaries and much of the "discipline" of the study. However, the wide range of sub-systems analysed and the relationships between them require that a wide range of analytical tools be employed. That is, a multi-disciplinary approach is required to effectively capture the reality, dynamics and diverse landscape traversed by the industry system.

Chapters 4 and 5 draw on secondary sources—official statistics and policy documents—to provide a broad overview of agricultural and industry development in China. The statistical analysis is based on close familiarity with the official statistics, is subject to critical examination and draws on additional analysis to derive unpublished figures. The chapters also draw on an array of policy documents to provide an "official" perspective on issues in Chinese agriculture and the beef industry and to act as context and cross reference to other parts of the study. Policy analysis is interwoven throughout the entire study.

Chapters 6 to 9 draw largely on primary information and provide a mainly qualitative examination of the industry sectors and the drivers and constraints to their modernisation, which can be technical, policy, socio-cultural or economic in nature. However, the analysis also draws on various forms of quantitative analysis, including consumer surveys (Chapter 6) and budget and sensitivity analysis of beef processors (Chapter 7) and cattle producers (Chapter 9). Some of the results of these analyses are also used to examine marketing systems (Chapter 8). The quantitative analysis is linked across chapters, particularly as input parameters for downstream actors in the agrifood chain (such as abattoirs) are coordinated with output parameters from upstream actors (such as cattle producers). This allows for quantitative analysis of relationships throughout agrifood chains, which span sectors and chapters of the study.

In addition to drawing on a range of analytical tools, the study also draws on a range of data. The data includes: official statistics and policy documents (from national down to local levels); extensive fieldwork and interview material (from nearly 500 face to face interviews); and micro-level survey and budget analysis. All three types of data have been collected and updated over a period of at least 10 years, and in more than 18 provinces of China. It is argued that rigorous cross-verification of these diverse data provides a robust and reliable picture the industry.

When considered individually, methods used within each chapter appear to be technically relatively straightforward. However, the integrative nature of the study can be demanding especially as integration occurs across disciplines, disparate data sources and across industry "sub-systems". While this approach poses particular research challenges, it is argued that it helps capture the increasingly complex reality and dynamics of the Chinese agriculture, livestock and beef sectors, to assess the implications for industry participants and to design appropriate policy responses. It also enables novel findings to be made that can contribute to broader understanding, as highlighted in the final recommendations and policy implications chapter of the book (Chapter 10).

CHAPTER TWO

LITERATURE REVIEW

To identify previous insights useful in addressing the study objectives, the literature review below overviews relevant work in the fields of rural development, the development of markets, agribusiness, smallholder systems, and change in Chinese livestock structures and supply chains. Relatively few studies examine these issues based on detailed examination of a particular industry, or conduct detailed budgeting of economic agents in the industries. The adoption of such an approach shows how agricultural and agribusiness development impacts on rural development, by establishing the returns, viability, options and mobility paths of different types of economic agents, both on-farm and off-farm. Compared to the patchwork of studies that do take this approach, this study differs in that it draws on a novel analytical framework which, among other features, adds the dimension of market and supply chain segmentation.

2.1 The role of agriculture in development

One of the underlying themes of this study is the role of agriculture in rural development. To summarise the enormous body of literature on the role of agriculture in development, a brief chronological overview of the major issues is provided. In the 1950s and 1960s, agriculture was largely ascribed as playing a "passive" role in national and rural development. Agriculture was treated as a small scale, traditional activity characterised by low productivity and decreasing returns that held few prospects for growth. Economic development was thought to involve a reallocation of resources from the agricultural sector to a modern industrial sector with higher productivity and increasing returns (Adelman, 2001). Agriculture thereby contributed only passively to economic development by providing surpluses of tax revenue, cheap food and labour. Where growth was to occur in the agricultural sector, it was to be undertaken by the large scale modern agricultural sector. As this took place alongside the traditional agricultural sector, a dual economy was formed (Fei and Ranis, 1961). This form of "modernist" thinking was also prevalent in socialist states like China and the Soviet Union with the centralisation of farming, and the formation of processing and marketing monopolies (Ellis and Biggs, 2001).

This passive view of the role of agriculture was reassessed in the late 1960s and early 1970s when the Green Revolution transformed traditional agriculture into a modern and potential growth sector (Diao *et al.*, 2006; Schultz, 1964). Johnston and Mellor (1961) argued against development based on the dual economy model and showed that small traditional farmers allocate resources efficiently given correct price incentives. The "rural growth based on small farm efficiency" or "small farm first" paradigm has since been a persistent theme in the rural development literature (Ellis and Biggs, 2001).

The more active role of agriculture in economic development was reinforced by recognition of the linkages between agriculture and other sectors, including backward linkages into inputs and services and forward linkages into agro-processing (Johnston and Mellor, 1961). In contrast to urban biased industrialisation characterised by dualistic development patterns, agriculture-industry linkages and interactions were argued to provide incomes for the poor both in agriculture and in agro-industry (Adelman, 1984; Singer, 1979). The importance of small farms in generating "rural growth linkages" or labour intensive off-farm activities in rural areas represents another persistent theme in the rural development literature (IFAD, 2001; Johnston and Kilby, 1975; Mellor, 1976). As in the case of China, these linkages deepen as an economy modernises but decline in relative importance alongside agriculture's share of production (Haggblade *et al.*, 1989). High profile leaders (Hu Jintao) and researchers (Ke, 2008) are now focusing on reverse transfers from industry to agriculture (*gongye fanbu nongye*). In Chinese terms, after a modern history of agriculture "feeding" industry, the "baby bird" of industry has grown up to be much bigger than its "agricultural" parents and now industry must now fill the role of "feeding it's elders" (agriculture) through linkages and transfers.

Inter-sectoral linkages also featured in the debate on the role of agriculture in rural development in the 1980s and 1990s (Hazell and Haggblade, 1991; Hazell and Roell, 1983). Based on an econometric analysis of data from the China Rural Household Survey between 1985 and 1990, Ravallion (2005) reports significant cross-sectoral effects in rural China, notably from cropping to other farm activities (forestry, animal husbandry and fishing) and between these latter activities and off-farm activities (industry, processing, transportation and handicrafts). While Ravallion found fewer signs of reverse linkages, others argue that the growth of agro-industry and off-farm employment (including remittances) can help finance the purchase of agricultural inputs to further increase productivity (Reardon *et al.*, 1994) and relax capital and risk constraints in farming (World Bank, 2007). Infrastructure development is an important aspect of building inter-sectoral linkages, while the formation of social capital also plays a role in increasing confidence and trust between farmers, input suppliers, processors and banks to initiate non-agricultural business and commercial agriculture (Irz *et al.*, 2001).

Poulton and Dorward (2003) discuss the role of small-holders in development in terms of the distinction between "growth drivers" and "growth supporters". "Growth drivers" are processes that provide an initial stimulus increasing income flows in an economy such as price or productivity increases in tradeable products. "Growth supporters" are processes which increase multiplier effects to give wider and more sustained growth impacts from the growth drivers. These include consumption linkages for products with high income elasticities. Production linkages arise where a "growth driver" demands locally produced, non-tradeable inputs (upstream linkages) or local processing (downstream linkages). As discussed in the section below, livestock and horticultural products are argued to have high income elasticities in poorer economies with strong upstream, downstream and consumption linkages (Dorward and Kydd, 2005). As also discussed in following sections, benefits to small-holders require not only the presence of potential growth drivers and supporters, but also exchange mechanisms that: a) transmit demand signals across different players in the economy; and b) enable these players to respond to these with increased local supply.

Many of these themes are revisited by the World Bank (2007) in its 2008 Development Report titled "Agriculture for Development". The report argues that the world is in the midst of a "new agriculture" underpinned by new structures[3] and a series of "revolutions".[4] This has opened up new opportunities for rural development, arguing that "pathways out of

---

[3] "Dynamic new markets, far reaching technological and institutional innovations, and new roles for the state, the private sector, and civil society all characterize the new context for agriculture. The emerging new agriculture is led by private entrepreneurs in extensive value chains linking producers to consumers and including many entrepreneurial smallholders supported by their organizations" (World Bank, 2007, p.8).

[4] Successors to the industrial revolution and the green revolution cited by the (World Bank, 2007), include the agribusiness revolution, high value agriculture revolution, livestock revolution, horticulture revolution, aquaculture revolution, supermarket revolution, small-holder revolution, small-holder productivity revolution, bio-technology revolution, microfinance revolution, productivity revolution, while biofuels are a "revolution in the making".

poverty open to the rural poor by agriculture include smallholder farming and animal husbandry, employment in the new agriculture of high value products, and entrepreneurship and jobs in the emerging rural, non-farm economy" (World Bank, 2007, p.1).

The World Bank report uses a typology based on the contribution of agriculture to growth to distinguish between three different types of agricultural countries, namely: agriculture based; transforming; and urban based. China falls into the "transforming" category along with countries like India and Indonesia, where agriculture is a small part of the economy and where poverty is still overwhelmingly rural. The main challenge to transforming countries, especially China, is to address growing inequalities. As discussed below (Section 2.4.3), this is to be achieved through a diversified development strategy of traditional, small scale, welfare type agricultural activities, the pursuit of higher value agriculture, and the development of off-farm rural labour and migration.

2.2 Agricultural markets, agribusiness and farm structures

This section overviews literature related to the "high value agriculture revolution", the "agribusiness revolution" and farm structures. The review acts as a base on which more detailed case study specific analysis is undertaken in the body of the book, especially Chapters 6 to 9.

2.2.1 The "high value agriculture revolution"

In developing countries, agricultural produce in final markets is typically low priced, which is reflected in small margins for actors along the supply chains and ultimately in low returns for producers. Increasing consumer incomes and urbanisation in developing countries (Rae, 1998) as well as increasing agricultural exports in countries such as China (Nie, 2006),[5] have generated what has been called the "high value agriculture revolution" and is a source of considerable optimism for policy makers and development agencies (World Bank, 2007). Higher market prices may increase margins along the supply chain, increase opportunities for processing and other

[5] China's agricultural exports grew at 8.5% between 1995 and 2005, but growth rates are expected to decline to 7% for the period of the 11th Five Year Plan (2006 to 2010) to reach US$38 billion (Nie, 2006).

"value adding" activities and has the potential to increase revenues for producers that supply the higher value markets (Weinberger and Lumpkin, 2005).

A number of fundamental issues arise in the discussion about the "high value agriculture revolution". Most fundamentally, "high value agriculture" is often equated with a shift away from staple foods (such as grains and roots) toward non-staple foods (especially livestock, horticulture and aquaculture). These latter products generally have high income elasticities (hence growing demand), are perishable (and therefore risky), and where price sensitivity to quality and food safety is particularly important (Minot, 2005). There are ambiguities in applying the term "high value" agricultural products to broad classes of products. Non-staple agricultural products in developing countries are still dominated by mass markets. Furthermore, if "value" is equated with returns, analysis needs to take into account not only prices and revenues, but also costs and risks, which can be higher for the production, processing and transaction of non-staple agricultural products. In accordance, Minot (2005) defines high value agricultural commodities as those that generate high returns per unit of labour, land, or capital.

This study incorporates an additional level of specificity in the examination of high value agriculture. "High value agriculture" is seen in this study as a process of market segmentation—including within a single class of product—and the relative growth of higher priced segments of that commodity market. The study identifies three beef market segments (low, mid and high), which are linked to three beef agrifood chains (low, mid and high). The returns to the participants in those different market segments and agrifood chains are a function not only of revenues but also of costs and other factors, estimated through budget analysis. This form of high value agriculture requires a marketing system that is able to facilitate price-grade differentials, as discussed below. Findings of this study suggest that a cautious and critical approach to the "high value agriculture revolution" is required, as noted in other studies.[6]

There are very few studies that differentiate between market segments, partly due to the lack of available data. For example, official price data

[6] Diao and Hazell (2004) argue in the case of Africa that while there are opportunities in the traditional and non-traditional exports, they are growing from a small base and the greatest market potential for most African farmers still lies in domestic and regional markets for food staples. Minot (2005) concludes that high value agriculture—and relatedly contract agriculture—only makes sense for perishable quality sensitive goods purchased by processors, exporters or other large buyers.

routinely collected by the Ministry of Agriculture and reported in various yearbooks, newspapers and websites represents an "average" price for a "generic" product. Gale and Huang (2007) use official consumption and expenditure data, and hence per unit expenditures, to derive "quality" of food consumption (which was found to be increasing on average), but which does not account for different market segments. Wang *et al.* (2008) identified food safety characteristics as a modest source of premiums for milk which is supported by surveys reported for beef (Longworth *et al.*, 2001, Chapter 13). However studies based on consumer surveys rarely identify different markets segments for livestock products, including beef, as discussed in more detail in Chapter 6.

The disaggregation of value markets and segments is particularly important to industry. Industry bodies and companies constantly, albeit often informally, collect such information through for example industry meetings, observation of and participation in trade, and price surveillance. This study draws on similar types of "industry intelligence" of sufficient detail to enable the identification of market segments, as reported in Chapter 6. The approach is rarely used in the academic literature, and may contribute to further development of the market segmentation approach.

2.2.2 The "agribusiness revolution"

For the purposes of this study, "agribusiness" refers to all off-farm agricultural activities, both large and small scale (FAO, 2007a). On the inputs side, agribusiness includes agricultural inputs and services (although these are not the focus in this study). On the outputs side, the agribusiness sector includes retailing, processing and marketing (that are examined in detail in this study). The agribusiness sector transmits product and signals between consumers and producers. The efficiency and capacity of the agribusiness sector to do so have important implications for rural actors participating in and supplying into supply chains (Poulton and Dorward, 2003).

The pace of change in the agribusiness sector and the extent of the implications have generated what has been described as an "agribusiness revolution" (World Bank, 2007, p.135), with most attention afforded to the development of "modern supply chains". While the Chinese agribusiness sector is under rapid development and change, there are remarkably few China specific references to draw on and even fewer that analyse development implications.

Processors

The relative paucity of studies on Chinese agribusiness is particularly pronounced for the agro-processing sector. While there are several macro-level studies on rural industry (Byrd and Lin, 1990) and agro-processing (OECD, 2000), there are very few based on information from agro-processing enterprises themselves. This may be because of difficulties in collecting data. Examples of studies based on interviews and economic information from agro-processing enterprises in China include Brown, Waldron and Longworth (2005) and Brown, Waldron, Longworth *et al.* (2005) in the case of wool processing and, in much more limited way, for pigs in Fabiosa *et al.* (2005). Based on over 40 interviews with cattle slaughtering and beef processing operations, Longworth *et al.* (2001, Chapter 9) constructed representative budgets for four types of slaughter units: joint venture premium market abattoirs; modern (agro-industrial) abattoirs; county level and General Food Company abattoirs; and slaughter households. This latter category is significant given the dearth of studies on small scale and household processing, despite its dominance in many agricultural industries in China. Some studies (Hu *et al.*, 2004; Reardon *et al.*, 2001) state that smaller processing firms are left out of modern supply chains, with medium size and large processors preferred for long term contracts, although this is not the case in the beef and many other agricultural industries in China.

Marketing

A range of studies examine the reform of China's marketing system for agriculture as a whole (Lin, 1992; Wang and Davis, 1992; Watson, 1988), for the livestock sector in particular (Pan, 1998; Waldron *et al.*, 2003a) and for particular industries including pork and vegetables (Yu and Ke, 1992) and wool (Longworth and Brown, 1995). Several studies look at the dismantling of State marketing cooperatives including Supply and Marketing Cooperatives (Longworth and Brown, 1995; Sicular, 1996) and General Food Companies Waldron *et al.* (2003, Chapter 2). Much of the attention has focused on rural market places (Rozelle *et al.*, 2003; Skinner, 1964; Skinner, 1965a; Skinner, 1965b; USDA, 2005). Spot markets and millions of dealers and agents dominate virtually all aspects of agricultural marketing in China. Despite this dominance, few studies examine the operation of these dealers and agents, notable exceptions being Brown,

Waldron and Longworth (2005, Chapter 7) in case of wool dealers and Gong (2007, Chapter 7) for beef stall holders. Attention is now turning to the development of "modern retailing" especially the so-called "supermarket revolution" in studies such as Hu *et al.* (2004) and more critical and circumspect reports (Goldman, 2000; Goldman and Vanhonacker, 2006). While some studies cite the development of modern supply chains in China, this does not reconcile with the low levels of penetration at the producer level in livestock industries, with the notable exception of poultry (Bi, 2007; Gong, 2007; Guo *et al.*, 2007; Shen *et al.*, 2005; Huang *et al.*, 2008; Wang *et al.*, 2008).

An important aspect of market efficiency or how well the various market structures perform is to examine the integration of markets. Price integration analysis has been conducted on some grains (Huang and Rozelle, 2006), vegetables (Huang *et al.*, 2007), beef (Brown *et al.*, 2002b) and wool (Brown, Waldron, Longworth *et al.*, 2005). The studies generally show increasing spatial market integration in China. However, a lack of data on price-quality differentials means that few studies examine the integration of markets by form of product, one exception for fine wool being Brown, Waldron, Longworth *et al.* (2005). Given the process of market segmentation and the need to establish price-grade differentials, this represents an important line of enquiry and a major gap in the literature. This study and related studies (Longworth and Brown, 1995; Longworth *et al.*, 2001) suggest that China's marketing system is highly undeveloped in facilitating price-grade differentials.

Explaining the underdevelopment of this aspect of the Chinese marketing systems can be aided by treating the marketing system as an institution. The trade in low value generic products can be conducted efficiently in a rudimentary market system governed by spot markets and inter-personal relationships (Lu, 2007). Li and Li (2000) see the development of market systems as a transition from "person based governance" (*renzhi*, where transactions are based on personal and implicit agreements) and "rule based governance" (*fazhi*, where transactions are not only observed by two parties but can be enforced by a third party). This involves market regulation, the replacement of subjective (sensory) measurement with objective (technical) measurement as the basis for describing product attributes, and the replacement of anecdotal information with verifiable public information as a basis for economic decision making. In this regard, the Chinese marketing system for the economy as a whole is undeveloped, especially in the areas of regulatory transparency and property rights.[7] Some of the specific instruments for normalising and developing Chinese market systems are discussed in Kumar (1994).

There are few studies that analyse the institutional dimensions—including the rules and systems that govern the way that production and transactions take place—of China's marketing system for agricultural products. Exceptions include Waldron *et al.* (2003a) and Waldron, Brown and Zhang (2007) which analyse "market facilitating policies" for beef in terms quality grading systems, food safety and hygiene standards, market regulation (registration of producers, traders and companies, and producer and consumer safeguards) and information systems (statistics, market reporting, market research). Some of these dimensions are also examined for wool in Longworth and Brown (1995) and Brown, Waldron and Longworth (2005). Rather than replicating the institutional dimensions of these previous studies by the author, the institutional dimensions are implicitly incorporated into analysis in the body of this book, and are briefly revisited in the recommendations and policy implications chapter.

Agro-industrialisation and associations

A concept closely related to agribusiness that has received an enormous amount of attention in China in recent years is "agro-industrialisation" and vertical integration. For a review of the Chinese "agro-industrialisation" literature in its early era of emergence, see Waldron (1999), and for a more recent review see Lingohr (2007). Amongst the enormous body of Chinese literature on Chinese agribusiness, one seminal study was especially important in quantifying the prevalence and characteristics of the agro-industrialisation movement in China. Niu and Xia (2000) reported on surveys conducted by the Ministry of Agriculture between

---

[7] The "Index of Economic Freedom" ranks China 126 out of 157 countries, where economic freedom includes an "absolute right to property ownership, fully realised freedoms of movement for labour, capital, and goods, and an absolute absence of coercion or constraint of economic liberty" . . . and "that freedom is both protected by the state and unconstrained by the state". For more information see http://www.heritage.org/research/features/index/index.cfm. Similarly, China was ranked 83 out of 181 countries in the "Ease of Doing Business" index of the World Bank/International Finance Corporation. The report "Comparing Regulation in 181 Economies" is available on www.doingbusiness.org.

1996 and 2002.[8] They report an increase of numbers of "agro-industrial organisations" or "dragon head" organisations from 12,000 in 1996 to 66,000 in 2000. The proportion of China's rural households that these dragon heads linked to increased from 10% to 25% in 2000. These high proportions arise because of the very liberal definition used of a "dragon head organisation", which was deemed to include agricultural enterprises, associations, and specialised market places. The Editorial Board of the China Township and Village Enterprise Yearbook (2006) reported that China had 580 national level dragon head enterprises in 2005 that linked with 87 million or 35% of the rural households in China.

Other studies that apply a more strict definition of a particular form of organisation find a much lower incidence of agro-industrial type organisation in China. In this regard, associations are an indicator as they are a common intermediary in contract agriculture and other forms of enterprise-farmer relationships. Shen *et al.* (2005) find that only 3% of farmers are involved in "real" Farmer Professional Associations, as opposed to associations that resemble enterprises or that are controlled by local officials. Chapter 8 of this study highlights the myriad of other forms of local groups in the livestock sector—especially "small livestock raising areas" and specialised villages—with varying levels of corporatisation and government participation.

Contract agriculture

Contract agriculture has been the subject of considerable attention from Chinese industry, government officials and researchers in recent years. Niu and Xia (2000) report that of the relationships between "dragon head organisations" and farmers mentioned above, 49% occurred through contracts in 2000. However, this proportion is lower than the estimated 71% in 1996. Of the contracts in 2000, only 43% were regarded as "genuine, signed, legally binding contracts" and the survey found that reneging on contracts by both farmers and agribusiness was commonplace. Based on a survey (220 households in various industries and areas in China), Guo *et al.* (2007) report a more positive picture of the state of

[8] For a more detailed summary of the study see Waldron *et al.* (2003, Chapter 10).

contract farming in China from the perspective of both households[9] and agricultural enterprises.[10]

Contract agriculture can be seen as an intermediate level institutional arrangement, lying somewhere between spot markets and vertically integrated commercial agricultural operations (Patrick, 2004). The development of contract agriculture has been depicted as a response to market failure, where small-holders can not participate in spot markets due to failures in the area of credit, insurance, information, inputs and product markets (Key and Runsten, 1999), or when price signals especially for quality characteristics are weak.

There are mixed assessments about the impact of contract agriculture and small-holders. Benefits include increase participation in markets, credit, technology, better risk management for small-holders, and increased employment opportunities. Proponents of contract agriculture cite higher income effects.[11] In the case of China, incomes for households with

---

[9] Of the farmers surveyed by Guo *et al.* (2007) only 21% (or 220) had entered into a contract (although more would like to do so if a contract was offered) mainly in order to increase market access but also for price protection. Most of the contracts were with dealers (which tended to be oral contracts) or firms (which tended to be written contracts), and marketing contracts were much more common than production contracts. Most of the contracts were based on a "flexible" (market) price basis, although farmers had a strong preference for a floor price system. Payments were made more on cash on delivery basis, rather than on cash after delivery or on deposits. 60% of farmers never experienced conflicts, while 36% experienced some conflict, mainly about price and quality. These were offset by the perceived benefits in the contracts, which were mainly in the areas of improving product quality, stabilising (but not increasing) sales prices, and reducing marketing costs.

[10] One hundred firms using contracts were also surveyed by Guo *et al.* (2007). Large firms tended to deal directly with producers, while smaller firms used intermediaries including villages, cooperatives, dealers and local government. Meat and dairy enterprises usually had production contracts, usually over a one to two year period, while vegetable and oil enterprises had marketing contracts, usually over a period of less than one year. Compared to smaller (local level) firms, larger (up to national level) firms tended to use written contracts with a floor price. 75% of the enterprises reported an acceptable contract performance, with contract failure attributed to unacceptable quality and contractees selling to higher bidders, especially by small sized produces, with most reporting "no solution" to conflicts. Contracts that use cooperatives as intermediaries perform the best, while those involving local officials perform the worst.

[11] Participation in modern supply chains increased farmer incomes in studies on peanuts in Senegal (Warning and Key, 2002), poultry in India (Ramaswami *et al.*, 2006), milk, broilers and vegetables in India (Birthal *et al.*, 2005), maize in Indonesia (Simmons *et al.*, 2005) and broilers in Indonesia (Patrick, 2004).

relationships with dragon head enterprises were claimed to be Rmb900[12] (Niu and Xia, 2000) to Rmb1,300 (Editorial Board of the China Township and Village Enterprise Yearbook, 2006) higher than those without the relationships.

Disadvantages for small-holders include the capture of smallholders within contracts with deteriorating contract terms, concerns about firm behaviour (Singh, 2000) and that contract systems are used to serve the interests of local elites (Guo *et al.*, 2007). Glover and Kusterer (1990) show only limited or negative benefits to small-holders. Another major cited problem is that the development of contract agriculture will benefit larger contracted households but exclude other smaller households thus causing a dual economy (Diao *et al.*, 2006; World Bank, 2007; Zhou, 2003). This latter issue is questioned in the case of the beef industry in this study.

Studies with an industry perspective on agribusiness

The studies overviewed above are generally cross-sectoral in nature and examine particular parts of the agribusiness sector (for example, rural market places, contract systems or supermarkets). While important, this approach can be usefully complemented with an industry or supply chain study that traverses all aspects of the agribusiness sector. In particular, an industry approach facilitates analysis of the relationships between different aspects of the agribusiness sector. Examples of studies that take this approach, include Longworth and Brown (1995) and Brown, Waldron and Longworth (2005) for wool; Longworth *et al.* (2001) and Gong (2007) for beef; Waldron, Brown, Longworth *et al.* (2007) for sheep meat; and brief overviews for pork are provided by Fabiosa *et al.* (2005) and Fuller *et al.* (2001).

2.2.3 Farm structures

The development of markets and the agribusiness sector hold important implications for primary producers. However, the heterogeneity of the primary production sector means that the impacts will be variable. There are various ways of ordering farm structures and their development. A typology of Hazell *et al.* (2006) orders farms along a continuum of: a) semi-subsistence farms; b) commercialising farms; c) and commercialised farms.

---

12 "Rmb" is used throughout the book to represent the Chinese currency unit Chinese Yuan Renminbi

They see agricultural transition as a process of small farms "stepping up" from being semi-subsistent to commercialising farms, and "stepping out" of agriculture into off-farm activities. Commercialised farms impact on the process because of their capacity to capture markets and out-compete smaller farms in some commodities and markets. The World Bank (2007, p. 75) use a different typology of: a) semi-subsistent farmers; b) market oriented smallholders; c) labour oriented households; d) migration oriented households; and e) diversified households.

While these typologies are useful and are referred to below, this study draws on the typology commonly used in China based around degree of specialisation and scale, namely: a) unspecialised households; b) specialised households; c) large "scale" producers. These boundaries are fuzzy in China and rural households pursue diverse and complementary livelihood strategies in not only farming but also labour and migration.[13] Literature on the non-farm sector is therefore also overviewed briefly below.

Unspecialised and specialised households

Unspecialised households—which can be equated with semi-subsistent households—allocate household resources across a range of activities, with partial integration with the market, especially low value segments of the market, through spot transactions. Such households play a valuable role in containing the spread of poverty, providing food security, retirement, identity, risk reduction and provide a "base" from which to consider other livelihood strategies (Hazell *et al.*, 2006). Livestock very often form part of unspecialised household systems and provide assets, food (milk, eggs, meat), manure, draught and transport power and cash surpluses (culls or offspring), and are often tended by women and children and can be held by poor households with very small land holdings not suited to cropping (FAO, 2007b; Hoffmann, 1999; Wilkes, 2003).

The role of small scale farming and livestock as a welfare type system in developing countries is highly applicable in China, especially in remote areas. However, there are real concerns that farm sizes and holdings in China are simply too small to be or remain viable in commercial terms (World Bank, 2007, p.92). This is particularly the case as growth in other parts of the Chinese economy increase the opportunity costs of labour and other resources in small

[13] For example, Knight and Song (2003) found that 65% of rural households and 33% of individuals in China operate in both the farm and non-farm sectors.

farms, which makes marginal small scale agricultural activities comparably even more unattractive, as discussed in Chapter 9 of this book.

There is also debate in the literature about how small semi-subsistent farm are being marginalised by commercialising and commercialised farms for a range of commercial reasons,[14] technical reasons[15] and "environmental" reasons[16]. Although data shows the continuing presence of small farms in developing countries, there is concern about the social ramifications of mass farm evacuation and urban migration and various measures have been recommended to help support small farmers (Hazell *et al.*, 2006).

Because of the very small size of farms in China and the opportunities in other parts of the economy, the debate in China is not so much about how to protect and support unviable small farmers, but how to help unspecialised farmers "step up" (within agriculture) and "step out" (of agriculture). Policy makers believe that some household farms can generate on-farm returns comparable to off-farm returns, provided that the farms become larger, more efficient and specialised. Chinese officials believe that this is best achieved by specialised farmers becoming the so-called "first workshop" in commercialised or industrialised commodity chains. These higher value chains can sometimes be distinct from those in which semi-subsistent unspecialised households are engaged, or there can be beneficial linkages. As a result, the relative importance of the two types of households need not necessarily be seen as a zero sum game.

Thus, the perceived problem in China is not that small farms are being replaced too quickly, but that they are not being replaced quickly enough. While some evidence (such as that presented for livestock below) suggests that livestock are increasingly being produced by larger households, this is happening from a low base and the pace of change has not been as rapid as might be expected. While regulatory reforms led to participation rates in

---

[14] De Haen *et al.* (2003) for example state that "small scale producers obtain lower financial profits per unit of output than large scale producers, other things equal. This suggests that, in the absence of deliberate action, small scale producers will eventually be put out of business by competition from large scale producers, especially since the better off producers will scale up". Delgado *et al.* (1999b) argue that policy distortions can magnify economies of scale in livestock production.

[15] Global and country studies (conducted in Brazil, India, Thailand and the Philippines) suggest access to breeding and feeding supplies and technology is critical to the success of poultry production, and is likely to become true for hogs over time (Pingali, 2007).

[16] In explaining the transition from semi-subsistent to commercialising farms, Hazell *et al.* (2006) cites factors such as increasing real labour costs, the development of markets for niche and high value products, the introduction of rental markets and the development of infrastructure and credit systems.

land rental markets of 10% in 2001 (Deininger and Jin, 2005), the increase in participation rates has since slowed substantially. A range of reasons for the ongoing importance of small farms are suggested in this study. For example, while younger generations pursue off-farm work opportunities, small farm and livestock systems still act as a welfare system for older generations, suggesting that systematic change may be generational.

Large scale producers

The next level up in the continuum from unspecialised and specialised household farm structures is what are termed in China has large scale—or simply "scale"—producers. Producers that fall into this category include feedlots and piggeries, large greenhouse projects, plantations and State farms. A point of differentiation between specialised households discussed above and "scale" commercial producers discussed below, is that the specialised household operations are usually owned and managed by the same person, where the owners of "scale" companies usually employ operational managers. In some cases, "scale" production structures are vertically integrated with processing operations.

Some industries and commodities are more suited to large scale, commercialised agriculture, especially where capital costs, input costs and risks are too high for smallholders, or where the costs of supervision of labour for menial labour tasks are low (Patrick, 2004). However, commercialised agricultural producers may not be competitive with smallholders in commodities where worker initiative is required, and where the commercialised operations represent high risks and high capital costs for investors (Hayami and Otsuka, 1993). This study suggests that labour costs are important factors in the relative competitive advantages of small and larger producers, especially when low opportunity costs for household labour on-farm are compared with wage labour off-farm.

Large scale livestock systems have become dominant in many livestock industries in both developed and developing countries.[17] In China, however,

[17] For example, 90% of pigs in the United States are produced by contractors with more than 50,000 pigs (Delgado *et al.*, 1999a). The four largest corporations—ConAgra, IBP, Excel, and National Beef—slaughtered about 84% of all US cattle in 2000 compared with only 21% in 1970 (Schlosser, 2001). In the mid 1990s, 80% of Thailand's poultry production came from only 10 large, vertically integrated enterprises that supply feed and day old chicks to medium and large scale producers under contract (Riethmuller, 2002).

with the exception of poultry, "scale" production systems are relatively unimportant in the livestock sector, with limited examples of intra-firm vertical integration. Rather than develop their own production units, agribusiness companies generally source inputs through contract systems (as discussed above) or—contrary to trends discussed in the literature—"open" spot markets. Nearly all large agribusiness agents (such as supermarkets, processors and even marketing agencies) interviewed by the author obtain at least some and usually most of their inputs through spot transactions on the open market (rural markets and traders) without contracts.

Relationships between farm and off-farm systems

As implied above, and explored more explicitly in this section, opportunities in the off-farm sector have a major impact on the incentives and structures of household producers. Indeed, the growth of the industry and services sectors in many developing countries has bought about what has been called a "rethinking of rural development" (Ashley and Maxwell, 2001). This school of thinking questions the role of agriculture as the engine of rural growth, and stresses the role of off-farm labour and migration in livelihood strategies and increasing rural incomes (Diao *et al.*, 2006). Ellis and Biggs (2001) and Ellis and Harris (2004) suggest that migration to urban areas should be facilitated to enable higher incomes and trickle down effects to agricultural areas where agricultural based incomes remain stagnant.

While the applicability of this perspective has been questioned in areas where there is a lack of off-farm opportunities or a lack of financial, social and human capital (Barrett *et al.*, 2001; Diao and Hazell, 2004), it is extremely applicable in many areas of China due to the high availability of off-farm employment opportunities. An estimated 110 to 130 million rural Chinese are working off-farm, with the vast majority of young rural people working off-farm (Rozelle, 2008). The impact of rising opportunity costs of labour on agricultural activities like cattle production, are examined in Chapter 9 of this book.

The incidence of off-farm work in China is also relevant to this study insofar as some of this off-farm work is conducted in downstream sectors of agricultural industries, including in the beef industry. In regard to the latter point, it is useful to consider the off-farm employment options for rural households, which include local agricultural employment, local non-agricultural employment, self employment and (temporary) migration. Shi *et al.* (2007) find in the case of Jiangxi that migration is the most important type of off-farm employment, but that migrant remittances are smaller than household incomes earned from self employment and local non-farm employment, while agricultural wage employment is a minor activity. These results were broadly consistent with those of De Brauw *et al.* (2002).

Despite the lack of data, it is reasonable to assume that much of the employment in local rural industry and services sector would be related to agribusiness activities such as agricultural processing, trading and transport.[18] Furthermore, much of the local agribusiness employment is not wage labour, but self employed enterprise amongst individuals, small companies, collectives and kinship and family groups. Off-farm activity is usually conducted by some members of the household in parallel with on-farm work by other household members, or both.

This study may contribute in a modest way to understanding the relationships between on-farm and local rural off-farm activities. The contribution is limited in that it only assesses options in the context of a single industry. However, fieldwork revealed many cases where entrepreneurial rural households "mobilise" and move between different types of activities within the same industry (e.g. cattle production, trading and slaughter) in which they had developed assets, expertise and contacts.

2.3 Livestock and industry studies

The review above refers primarily to the agricultural sector as a whole, but given the cattle and beef focus of this study, this section concentrates more narrowly on studies on the livestock sector and particular industries.

2.3.1 The livestock revolution

Delgado *et al.* (1999b) outlined and projected the enormous growth in the livestock sector to identify what was dubbed the international "livestock revolution".[19] This livestock revolution has occurred almost exclusively in

---

[18] Of China's rural working population, approximately 300 million work in agriculture, 60 million work in industry, 14 million in transport and 21 million in wholesaling and retailing (Editorial Board of China Agriculture Yearbook, 2006). A significant component of the latter three categories could be expected to be related to agricultural products.

[19] Simpson *et al.* (1994), Simpson (2003) and USDA (2007) also make projections on livestock growth in Asia and China.

developing countries, and in almost all livestock industries has been driven by developments in China (Delgado *et al.*, 1999; Waldron, Brown, Longworth *et al.*, 2007, Chapter 2). Delgado argues that the magnitude of the growth of the livestock sector is greater than the better known "Green Revolution". Like the Green Revolution, there are major implications, including that for the environment, health and development. These implications are not addressed in any detail in the report, but referenced to a series of country/industry case studies.[20] Given that the "livestock revolution" is driven largely by developments in China, the incorporation of a China case study would have enabled a more complete analysis. For example, Delgado argues that unlike the "Green Revolution" which was driven by supply-side innovations, the "livestock revolution" has been driven by demand-side factors. The analysis in this study argues that the drivers—and the constraints—have as much to do with supply-side factors and policies as with demand-side factors.

Largely building on the "Livestock Revolution" study of Delgado (1999), there have been a series of other studies by multilateral organisations on livestock and development, with a particular emphasis on livestock policies. Each of the studies identifies policy areas and proposes "policy frameworks". World Bank (2001) addressed livestock policies in financial terms, with an emphasis on liberalisation policies.[21] The FAO (2007b) were primarily interested in livestock policies that reduce poverty.[22] The FAO (2001) summarised policy issues that were of most concern to Asian policy makers, which were based on the "traditional" categories of production, marketing, infrastructure, political (e.g. farmer organisations) and the environment.

---

[20] These are on broilers and swine in the Philippines, broilers and eggs in India, swine, layers, broilers and milk in Thailand, swine, layers, broilers and milk in Brasil. These cases were also used in related projects: Livestock Industrialization, Trade and Social-Health-Environment Impacts in Developing Countries; True and False Economies of Scale in Livestock Production Using Concentrate Feeds; Causes and Consequences of the Industrialization of Livestock Production.

[21] The World Bank (2001) policy framework included the components of financial policies, international competitiveness, internalising environmental costs, introducing cost recovery for livestock services and institutional policies (for pastoral access, and for processing and marketing).

[22] FAO (2007b) propose a method for "Pro-poor review and analysis of livestock related policies". This includes grouping policies into three categories: policies that reduce vulnerability (land, water, feed, risk coping); policies that create conditions for growth (services, credit, markets); and policies for sustaining growth (food safety, competitiveness and the environment). Scoping the commitment of national and sub-regional government to livestock is also important.

2.3.2 The Chinese livestock sector

In reviewing studies on the Chinese livestock sector relevant to this study, three broad categories of studies are used below. First are survey-based studies that describe change in the scale and structures of the Chinese livestock sector. The second are budgeting-based studies that examine economic incentives as a cause of change. Third are studies that seek to explain change by looking at the broader supply chain and policy settings in which households act.

Survey-based studies on scale and livestock structures

One of the major indicators of change in livestock production structures and implications for small-holders is scale of production data, which is analysed in China primarily through official scale data of the Ministry of Agriculture (see Section 5.4.2 for an overview of the data). An analysis of this official data has been undertaken for pigs, broilers and layers Zhang (2006a), cattle (Longworth *et al.*, 2001; Waldron, Brown and Zhang, 2007), sheep (Waldron, Brown, Longworth *et al.*, 2007) and all livestock (Yan *et al.*, 2006). There are significant differences between types of livestock and livestock producers. For example, poultry production occurs on a much larger scale than other types of livestock. However, there are also common patterns across livestock types. While small households dominate all forms of livestock production, they are steadily decreasing in relative importance. This is concurrent with an increase in the relative importance of larger specialised household production systems. However, the large "scale" producers have shown only gradual or erratic patterns of growth, and often exhibit declines.

Another series of studies draw on survey data of the fixed point rural survey system of the Research Centre for the Rural Economy to analyse changes in household livestock production. Chen *et al.* (2005) uses the data to test the relationship between household pig production and the development of markets for labour (indicated through the percentage of the village labour force working in non-agricultural sectors) and grain (indicated by the percentage of grain marketed off-farm). They find an inverted U-shape relationship where households increase pig production as access to markets increase until a certain point when pig production then decreases due to the increased opportunity costs of labour and feed. As labour and grain markets are more developed in Eastern China, market development is used as an explanation for the Westward movement of pig production in China.

Rae (2008) and Rae and Zhang (2009) use the same survey data to examine the relationships between household livestock specialisation and incomes between 1995 and 2005. They show that households that earn a small percentage of their income from livestock are most likely to exit livestock production. This occurs alongside increasing specialisation among the households that remain in livestock production and which forms an increasing gap between small and large households. Like Chen *et al.* (2005), they also identify the relative shift in livestock production toward Western China, with the exception of broilers and layers. They predict a U-shaped relationship between incomes and specialisation in livestock especially in Eastern China, where household incomes were highest with low levels of specialisation and where incomes are high with high levels of specialisation in livestock. It was conjectured that the correlation between specialisation and incomes occurred most strongly in Eastern China because of access to modern supply chains, and questions were raised about the social desirability of the trend.

Based on an independent survey, Bi (2007) examined the marketing channels of poultry and pig producers around Beijing. Bi (2007) found that pig producers continued to market outputs through "traditional" marketing channels (dealers and local slaughter units); that poor small-holder households have increased their relative importance in the pig production sector; and furthermore that these smaller households have increased their average number of pigs in stock. On the basis of these results, Bi (2007) concludes that traditional market channels for pigs are "pro-poor". In agreement with a related study on horticulture (Huang *et al.*, 2007), Bi (2007) found that traditional channels remain competitive because of undeveloped transport and other infrastructure, undeveloped cooperative structures, and price sensitive markets. As could be expected, Bi (2007) found that a much higher percentage (62%) of poultry producers sold directly to "modern supply chains" (especially processors) and sourced inputs (feed and veterinary services) from the modern supply chains, and only enabled participation from rich households. The findings for poultry are said to support concerns that poor households will be excluded from modern supply chains.

These studies are useful in describing broad trends in household livestock production and provide useful empirical evidence of "what" is happening. Furthermore, the findings are broadly complementary and conform to many of the findings reported in this study. However, survey-based studies can be less instructive on "why" the phenomenon has occurred if they lack data on which to derive explanatory causes. For example, many survey-based livestock studies do not distinguish between household breeding operations and feeding operations, which are very different and these differences are likely to impact heavily on the findings.

Budgeting-based studies of livestock production systems

Budget analysis of livestock production systems can provide a more disaggregated perspective of why livestock particular systems exist and change. In particular, such analyses can help identify the stimuli and incentives for households to enter or exit livestock production, change production systems, and respond to price, technical, policy and other signals. The capacity of budgeting to provide an accurate and relevant picture of these economic incentives, however, depends on the representativeness of the budgeted production systems, the capacity of the budgeting to examine various system changes, and the quality of the data.

Especially since the 1990s, a relatively large number of studies have budgeted various aspects of livestock production in China. Two sources of data are commonly used based on large official surveys. One is the survey data of the Research Centre for the Rural Economy mentioned above which is used for example by the National Office for Rural Household Surveys (1998), Zhang (1998) and Zhang and Lu (1997). While useful to analyse scale and structures, the data is too highly aggregated for detailed budget analysis of variable household production systems, breed and feed regimes and marketing channels.

Another "official" data set that is more disaggregated is that of the "National Agricultural Products Cost and Revenues Material Collection" collected and reported by the Commodity Price Department (of the National Development and Reform Commission) in collaboration with the Ministry of Agriculture. Yan (2006) provides a recent and comprehensive use of this data to analyse change in the livestock sector. Zhang (2006a) uses the data to budget pig and broiler production in unspecialised, specialised and corporate production systems. Waldron, Brown, Longworth *et al.* (2007) use the data to budget: a) household meat sheep and goat production in intensive agricultural systems; b) specialised household local breed sheep production in extensive pastoral systems; and c) specialised household improved breed sheep production in extensive pastoral systems.

The Commodity Price Department data disaggregates between various types of costs and revenues to generate household net profits and cost-revenue ratios, to enable comparison across types of producers. The advantages of the surveys include that they allow for comparison across regions and a long continuous time period. The drawbacks of the surveys are that data is averaged out to provincial levels, which cover an enormous range of systems (even within the category for example "beef cattle production in intensive systems"). Critical parameters such as production system (e.g. breeding versus fattening), time on feed, growth rates for different feed and breed types, and specific prices are not surveyed or reported.

More tailored and disaggregated budgeting approaches have been used to assess the economics of beef cattle production based on crop residues in Wang and Zhang (1994) and Liu (1998), and for feed use mainly for pigs in Fuller *et al.* (2001). More disaggregated and technically detailed surveys were conducted for specific types of beef cattle production systems (unspecialised cow–calf, specialised feeding and feedlot operations) in Longworth *et al.* (2001) and form the basis of the budgeting of producers in this study.

Supply chain studies

A number of studies have sought to integrate household level budget analysis with a qualitative analysis of developments that occur along the supply chain. This supply chain approach can contribute to an understanding of impacts and incentives at household level. Fuller *et al.* (2001) supplemented their household budgeting with a section on "interviews and anecdotes" including on marketing channels and enterprise sales channels. Fabiosa *et al.* (2005) budget the operations of some parts of a commercial pork supply chain in China, namely a commercial piggery, two slaughter plants and two retailers. This is used to establish price spreads through the supply chain. The distribution of retail value is much more highly concentrated with pig producers in China compared to the United States where value adding means a high distribution between wholesale and retail segments. Zhang (2006a) supplemented his "cost-revenue analysis" budgeting with a small survey of pig and dairy producers to assess their source of inputs (including information, training, capital, insurance, breed services, pens, land) and outputs (markets, enterprises).

A series of short reports commissioned by the Ministry of Agriculture look specifically at the issue of increasing livestock prices in 2008 and the impact that this has had on profits along the supply chain, especially producers. These are conducted for pork (Nan and Han, 2008), broilers (Zhai and Han, 2008) and dairy (Li *et al.*, 2008). The short reports, lack of information about data collection and analysis, and the summarised reporting of the budget data make it difficult to comment on the analyses. However, each study concluded that the increases in prices for livestock products (pork, chicken meat and dairy products) in 2008 did not necessarily result in increased returns to producers because of the corresponding increase in feed and other costs of production, and because most of the increase in returns were extracted by downstream stages of the supply chains (traders and retailers).

Longworth *et al.* (2001) combine budgets of different types of cattle producers with budgets of different types of slaughter operations. These budget analyses are set in the context of detailed chapters on other aspects of the industry including cattle marketing, beef marketing and beef consumption. Gong (2007) also surveyed cattle producers to assess their choice of marketing channels,[23] processors to assess their choice of procurement channels,[24] and dealers to assess their behaviour.[25] Transaction and negotiating costs were found to be a major determinant of choice of sourcing and marketing channels for cattle. From the analysis, Gong recommended that the beef industry as a whole build institutions and policies to increase levels of quality and safety, trust in the marketing system, and farmer cooperatives.

2.3.3 Industry studies with a policy and institutional focus

Another group of studies overviewed below focus more specifically on the political economy of agricultural industries with a focus on the role and

---

[23] Gong (2007) found that the choice of marketing channels for producers were determined mainly by three variables grouped as negotiations costs. Most producers generally choose to sell on open markets (markets or dealers) because of their low bargaining power and due to payment risks (non payment and delay). However, larger cattle producers (mainly feedlots) are more likely to sell directly to processors to reduce risks and increase prices.

[24] The choice of procurement channels by processors are also based on transaction cost considerations, the most important being supplier type, followed by trust levels, traceablity and capacity to monitor. Gong (2007) found that processors prefer to enter into contracts and least prefer to source cattle through open markets. Large scale processors are most concerned about continuity of supply than rather than capacity to trace back supplies (for food safety and quality).

[25] Markets remain by far the dominant channel for beef purchasers, either individuals or food retail outlets, while margins for beef retailing are low in all channels, including supermarkets. Beef dealers did not recognise any price-grade/quality differentials, either in the purchase of their carcass or sales to customers, and had a very low level of awareness of any quality assurance scheme or punitive measures for unscrupulous marketing.

performance of policy and the role of the State. With regard to the Chinese livestock sector as a whole, the policy analysis of Ke (2002) remains highly relevant today and is revisited in the final chapter of this book. At a more specific industry level, a group of studies use mainly qualitative methods and project experience to provide a micro level perspective of developments in the industries which can not always be captured through the reporting of survey based numerical data.

A study that is not directly related to livestock but which analyses some of the themes overviewed above is that of Lingohr (2007) titled "Rural households, Dragon Heads and Associations; a case study of sweet potato processing in Sichuan Province". Lingohr describes the nexus of components in the sweet potato supply chain at the local level[26] and the implications that this has on participating households.[27] As such, the study shows the way by which high level policy edicts can operate and impact in practice at local levels, with a mixed assessment in the case of development of agro-industrialisation and associations.

The role of the State is pursued more specifically by Delman (2003) in his article entitled "Cool Thinking? The role of the State in shaping China's dairy sector and its knowledge system". The article overviews developments in dairy production (farmers), processing, the extension system, and farmers' organisations. He outlines how the State at all levels

---

[26] Linghor (2007) overviews the growing literature and policy emphasis placed on the agro-industrialisation movement and farmer associations in China. She then applies this to case study on the sweet potato industry in Sichuan, which has various stages of primary production (sweet potato production), primary processing (starch) and secondary processing (such as noodle making). The study is based on interviews with two dragon head processors and two associations, and rural households and government officials linked to the organisations. Analysis is largely qualitative in nature but includes some gross revenue data from various industry activities.

[27] Rural households link with the dragon head enterprises through either employment in processing (primary or secondary), raw material production, or by leasing land and then providing labour to produce sweet potatoes. Production bases are established to produce sweet potatoes for the processors based on contracts between the secondary processing dragon head enterprises and township governments or primary processors. These contractees then organise the farmers to produce to meet the contracts. Output quantities, inputs from processors and floor prices are set. In some cases, households lease out their land to the contractees and provide labour for sweet potato production. Associations are established to organise and integrate the systems, and provide mechanisms for delivering services, market access and higher incomes. Amongst the drawbacks of the dragon head/ association system are that decision making powers and trust have been vested in local government, and there were suspicions of monopsonistic price setting, and the distribution of risk to producers.

have involved themselves in all these aspects of the industry and played an instrumental role in growing technical and organisational industry capacity. However, this interventionist "hands on" role has taken place at the expense of the development of what is termed a "knowledge system". He laments an absence of institutions that facilitate information flows, participatory problem solving and market based integration along the supply chain. This has major and long term impacts on the technical and allocative efficiency of the industry and ultimately the quality of dairy products, expressed most dramatically in the "melamine incident" of 2008.[28]

Delman uses the dairy industry as a case to assess the nature of State intervention in China. He draws on a categorisation of Blecher and Shue (2001), who distinguish between a development, predatory, entrepreneurial and clientist State. Blecher and Shue find that in the case of the leather industry in Xinji County in Hebei that the local State has become almost a "pure" development State. However Delman argues that in the case of the dairy industry the central State—especially the Ministry of Agriculture—has provided opportunity for local government to adopt all four types of intervention.

Waldron *et al.* (2003a) report similar findings in the case of the beef industry. They analyse in detail the organisation of the industry (including processors, associations, farmers, marketplaces), institutions (vertical and horizontal and other associations), and policies (industry policies, agro-industrialisation, market support policies, social policies and food safety policies). The book outlines the role of the State in proactively kick-starting a new industry to provide additional development options for many millions of rural households. Industry policies analysed resemble those pursued in an earlier era by the East Asian Newly Industrialised Countries (MacIntyre, 1989; Lindblom, 1977) through close government-business relations and corporatist structures (Unger and Chan, 1995; Oi, 1999).

While China has been largely successful in expediting the development of a new rural industry, this has occurred at the expense of over-investment and subsequent correction with significant costs. As the beef industry matures, the industry is now facing a "second round" of development which requires a new set of "facilitative" policies and institutions to bring about sustainable industry development. This represents a more challenging era

[28] Given the relatively high technical requirements in dairy cattle production, cold chain management and the relatively concentrated structure of the processing sector, this is of major consequence. In comparison, while beef is also a perishable product and technically demanding in premium markets, the majority of the beef industry operates under relatively low technical levels and is relatively localised.

than previously in which development will not happen "automatically" or through the more traditional top down approach. Subsequent analysis undertaken in more recent years in Waldron, Brown and Zhang (2007) reveals that this challenge has not yet been met. Instead, interventionist policy toward cattle producing households had been redirected toward large modern abattoirs as discussed further in Chapter 10.

There are both similarities and differences with the beef industry in the newly emerging sheep meat industry documented in Waldron, Brown, Longworth *et al.* (2007). The "production base" of the sheep meat industry was built extremely rapidly, but unlike beef, this occurred largely though private or quasi State sheep breeding enterprises. Rather than holding the sheep in centralised facilities, the breeding companies disseminate breeding sheep to households to take advantage of lower labour, capital and overhead costs. The households are organised into associations or "small livestock raising areas" and very often supported through government projects (especially poverty alleviation funding). Contractual relationships establish the proportion of offspring that are retained by the households and the proportion "returned" to the breeding company. Meat sheep are thought to be well suited to this type of project because they represent smaller (more divisible) units than cattle, and particular breeds and crosses have very high birth rates. The sheep "returned" to the breeding companies are sold on to similar projects in other areas, or on to the commercial sheep market which is thought to be a growth industry. The optimism stems largely from the popularity of hot pot cuisine, in which sprawling restaurant franchises are an important outlet and that are sometimes integrated back to slaughter stage. While the logic of this dual "policy-push"/"market-pull" development strategy can be discerned, there are signs that the sheep meat "bubble" is beginning to burst as the "domino effect" of the breeding projects breaks down, which will be accelerated if the popularity of hot pot passes as a culinary fashion. The goat meat industry went through a similar cycle a decade earlier.

The wool industry has a longer development history much of which has been documented in Longworth and Williamson (1993), Longworth and Brown (1995) and Brown, Waldron and Longworth (2005). The wool industry was targeted for development in much of western China, for both fine wool sheep production (including breeding) and processing, through mechanisms ranging from domestic trade barriers to price support schemes. The reform of these policy mechanisms of the central planning and early reform eras, and the marginal viability of growing or processing fine wool in Western China were the primary reasons for the decline of the industry in the 1990s. However, another contributing factor is that China has been unable to develop a fine wool marketing system that delivers product in a form required by processors, and that allows price-grade differentials to be passed back to producers. Like the beef industry, this is not due to a lack of industry standards issued, but due to a failure to provide the policy and institutional environment for the standards to be taken up (such as credit markets, price discovery, and market regulation.).

2.4 Development strategies and frameworks

Given the complex relationships between structures and issues mentioned in the literature review above, an integrative framework can be useful as a means of ordering analysis. A number of potentially applicable frameworks are overviewed below. While these provide useful reference, the study uses a customised framework described in more detail in Chapter 3 (Section 3.2.1).

2.4.1 Sustainable livelihoods framework

The sustainable livelihoods literature has become increasingly relevant in recent years, especially because it integrates off-farm activities as a determinant of rural livelihoods. Ellis (2000b) notes that a dictionary definition of a livelihood is a "means to a living" and emphasises the way by which livelihoods are derived. He modifies a framework of Chambers and Conway (1992) to derive the definition "a livelihood comprises the assets, activities, and access to these that together determine the living gained by the individual or household".

Figure 2.1, which is drawn from Ellis (2000a) and ILRI (2007), shows the components of these "three A's" and some of the important cause and effect relationships. This livelihoods framework is useful in framing the issues addressed in this study from the household perspective. Some of these issues are analysed in detail in this study, including:

- Assets (such as cattle, feed and labour)
- Access (institutions such as markets in practice; and organisations such as associations).
- Activities (natural resource activities such as livestock; and non natural resource activities such as rural trade and rural manufacture), and
- Livelihood security (mainly income; but incorporation of some income stability; and degree of risk factors).

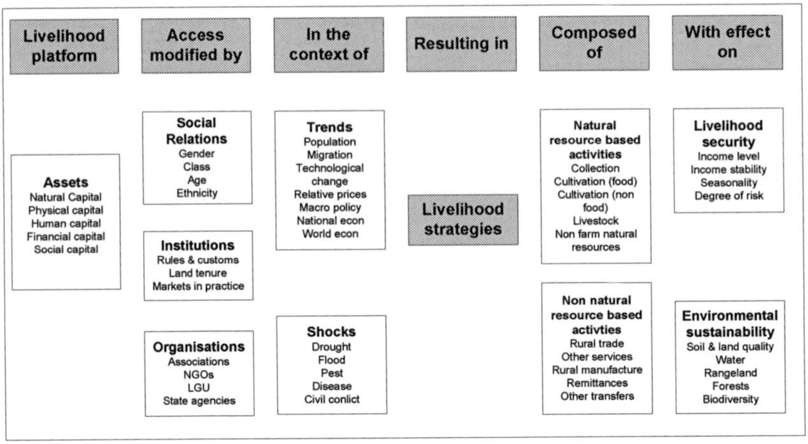

Figure 2.1: Components of the livelihoods framework
Source: Adapted from Ellis (2000a) and International Livestock Research Institute (2007)

One project that does use the livelihoods approach to order and link many of these factors is titled "Enhancing the Livelihoods of Agro-pastoralists in Northwest Yunnan".[29] The papers generated by the project examine complex relationships between diverse livestock systems, natural resource management, technical and service issues (such as disease control), community rules, relations and systems. On-farm activities and systems are mapped in detail and used as a basis to survey households (through participatory research) to rank the importance of various households resources and risks.

Some of these factors are analysed in this study, but many other factors are not systematically examined. This is partly because a single industry case study precludes analysis of the mix of activities from which a household derives a livelihood. Further, the livelihoods framework is not suited to application of a whole supply chain that incorporates downstream activities of the supply chain such as processing and final product consumption.

2.4.2 Value chain squares

Another framework that is more relevant to the role of households in off-farm parts of industry activity including markets and agrifood chains is

[29] For details see http://www.cbik.ac.cn/Get/English/Whatwed/Community Livelihoods/111049103.html

that of the Royal Tropical Institute (2006) entitled "Chain empowerment: Supporting African farmers to develop markets". The study differentiates between supply chains (characterised by loose coordination and often speculative conduct) and value chains (where each of the actors in this chain work together to service particular markets).

Farmers participate in value chains on two dimensions. On one (vertical) dimension are the type of activities that farmers undertake, and their degree of specialisation. On another (horizontal) dimension is the involvement of the farmer in the management of the chain. Combining these dimensions forms the matrix shown in Figure 2.2 comprised of four categories, but with fuzzy boundaries. Households and household groups can thereby act as one or more of the following actors:

- Chain actors which engage only in farming and have no influence over the management of the chain. Farmers in conventional contract farming schemes are also regarded as chain actors.
- Chain activity integrators which have moved from farming into other activities such as milling in the chain, yet without exerting more influence on the management of the chain
- Chain partners that also specialise in farming but also involved in associations or groups to exert influence over the management of the chain, and establish long term chain partnership with traders, processors or retailers and input/service providers.
- Chain co-owners which have moved upstream in the chain, increasing both their activities and their influence, and are organised in business cooperatives that develop new products and reach the end consumer

Most farmers are chain actors located in the bottom left hand corner of the matrix. Farmers can "empower themselves" by moving up and to the right into other categories, although the authors of the study note that there is no predetermined "better" position. Farmers or groups also move in a multi-dimensional way between boxes depending on incentives and developments in the value chains. The study draws on a large number of brief descriptive qualitative case studies.

The framework provides a compact illustration of the options of households in production, marketing, processing and even retailing activities. The broader study also raises important issues on the organisation of farmers, power relations with other actors in the value chains, and improvements in the operation of value chains through cooperation. All of these factors are important in the Chinese context. The framework is designed to be flexible enough to use in a large number of scenarios and therefore lacks the specificity inherent in the framework used in this study.

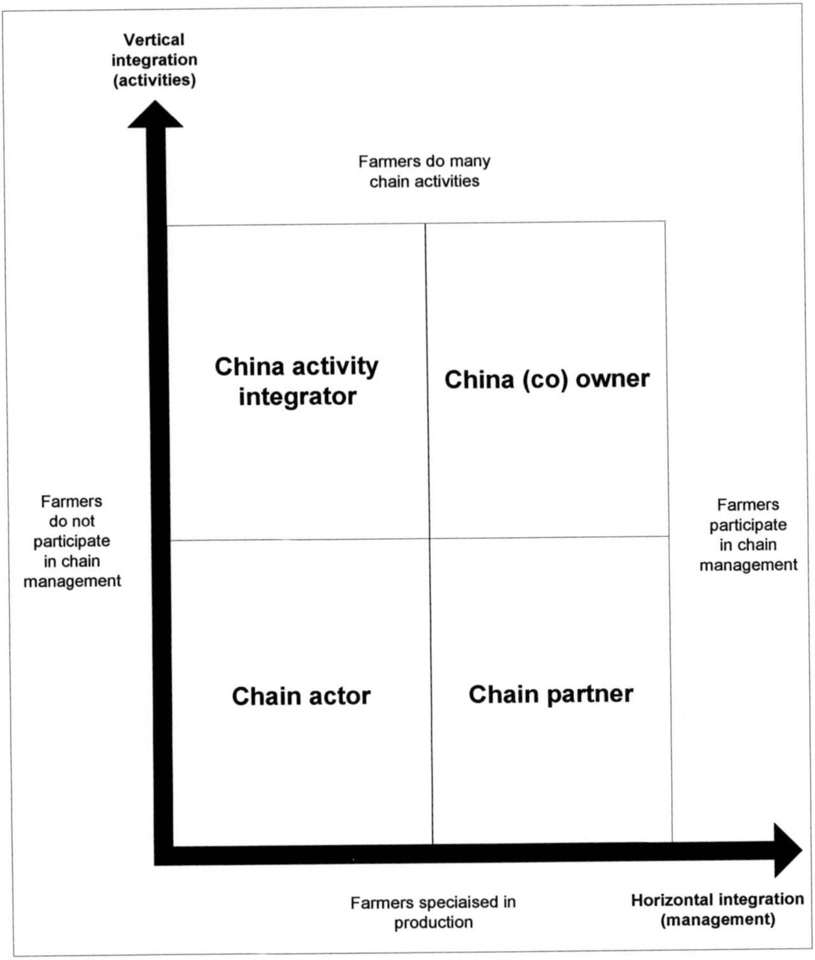

Figure 2.2: Forms of chain participation by small farmers
Source: Royal Tropical Institute (2006)

2.4.2 World Bank policy diamond

Another useful framework with a broader focus that those described above is set out in World Bank (2007). The report first proposes a "policy diamond", shown in Figure 2.3, to set the agenda for developing countries to open up "pathways from poverty". It then goes on to prescribe policy agendas for different categories of countries, including transforming countries such as China, as shown in Figure 2.4. These illustrations also act as a compact summary of some issues addressed in this study. However, the specific issues addressed in the study are ordered around a more detailed "industry system" framework. Rather than just being used as an aid to conceptualise the issues, the framework is attached to or is designed to guides specific types of analysis.

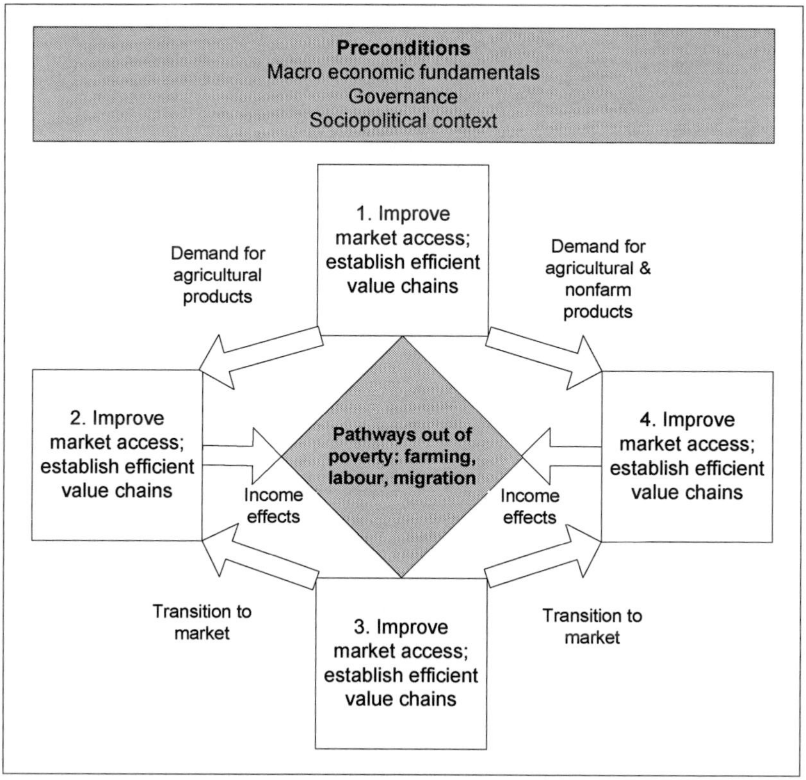

Figure 2.3: The four policy objectives of the agriculture-for-development agenda
Source: World Bank (2007, Figure 9)

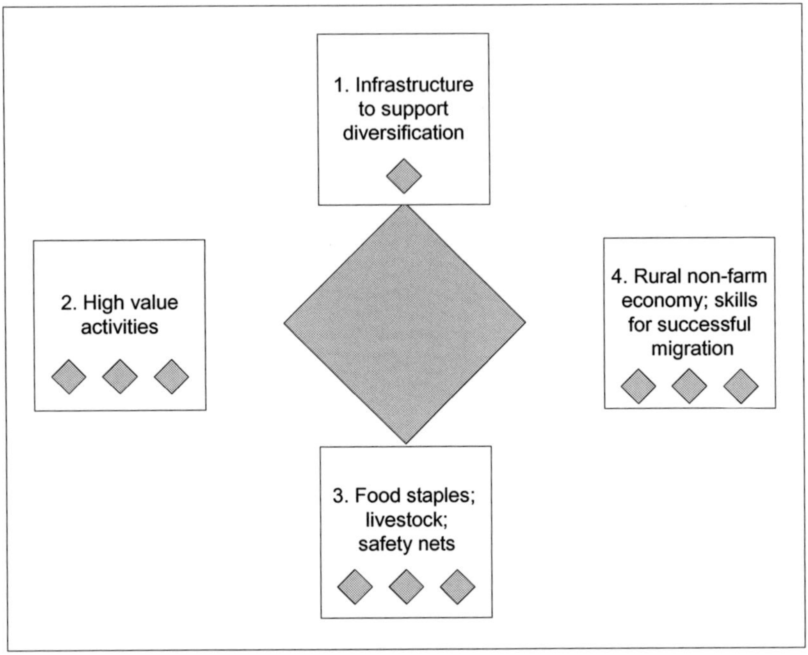

Figure 2.4: Policy diamond for transforming countries
Source: World Bank (2007, Figure 10.2). Note the number of diamonds the relative priority of policy attention, assigning 10 points across 10 objectives

Chapter Three

Methodology

As already mentioned in the introductory chapter, the study is based on a detailed industry case study approach. The beef industry is disaggregated by sectors and by value chains to form an "industry system" comprised of a series of sub systems. As such, the study takes a systems approach rather than a reductionist approach to the research. The industry system and the relationships between the sub systems are analysed through multiple analytical tools and sources of data, namely: statistical data and policy documents; survey data and budgets; and a large amount of interview data collected over more than 10 years of fieldwork. The fieldwork in particular has allowed for an inductive rather than deductive reasoning approach to the research. It is argued that the cross-verification of the multiple forms of research data and findings provides a rigorous, in-depth and "grounded" form of analysis that can aid understanding of China's increasingly complex and fast changing agricultural and rural economy.

3.1 An industry approach

Industry studies are rarely used to investigate broader issues in agricultural and rural development. While the discussion below highlights some of the reasons why this might be the case, the value of using the industry approach is also presented. It is argued that the industry case study can add nuance and veracity to the literature and debate on agricultural modernisation as well as act as a point of comparison with other industry studies. This is in addition to the applied specific contribution to understanding the beef industry in China, which is important in its own right.

3.1.1 Insights into agricultural and rural development

Industries or commodity chains provide a window into broader economic phenomenon. The analysis of industry supply chains allow for examination of economic agents and their relations, including material flow (various products and levels of transformation), method of transfer (market, contract), returns and rents extracted along the chain (labour, investment), organisation (household, associations, companies) and technologies. "External" environmental (policies, institutions) impacts and power relationships along the supply chain can also be examined. While these factors are organised around a particular commodity (or group of commodities) or supply chain, the aggregation of these chains can be seen to form an economy (Rammohan and Sundaresan, 2003).

As implied by the wide scope of topics covered in the literature review in Chapter 2, an agricultural industry traverses a broad cross section of the agricultural and rural landscape including markets, processing, marketing and production, extension and services, policies and institutions. An industry analysis therefore allows for an examination of the relationships between different aspects of the agricultural sector. An industry focus can also be seen as a meso-level of analysis that links the macro-level (policy, institutional and the broader economic environment) with the micro-level (household and other economic agents). The addition of this meso-level analysis adds a richer and more accurate understanding of the causal explanations (the "why" and "how") of how macro settings impact at the micro-level. In other words, the addition of an industry dimension can add to understanding of the mechanics of the process of development.

This in turn can allow for more targeted policy responses. As the Chinese agricultural economy becomes increasingly sophisticated, and the easy and obvious solutions to developing Chinese agriculture and development become exhausted, other more nuanced and incremental changes will become increasing important. Thus developing appropriate policy responses to the more complex environment requires understanding of the detail and reality at a micro-level.

The industry approach, however, also involves limitations. Within the series of agricultural industries that comprise the agricultural sector, analysis is centred on one particular industry. Analysis is therefore unable to capture the choices that economic agents make between alternative industry activities or mixes of activity. For example, decisions and mobility paths of rural household producers are usually not confined to one industry but span or extend beyond agricultural industries. The decisions of downstream actors can be even less confined to the bounds of the industry if they invest horizontally in other industries.

While the limits of the industry approach are acknowledged, some effort is made to incorporate findings from other related industries. This is especially the case with other ruminant livestock industries on which the author has comparable prior knowledge (wool, sheep meat, and to a lesser extent dairy). In some parts of China and in some ethnic minority areas, economic activity and decisions can be confined to these ruminant livestock sector. While this suggests that a sectoral approach rather than industry approach would have been more fruitful, it proved unwieldy using the framework and methods discussed below.

3.1.2 The case study approach

Case studies are a form of empirical research that examines a situation as it exists in real life (Yin, 1994). Case studies provide for compelling construction of theory and the research is likely to be considered more robust. Multiple cases can provide for modification of emergent theory, provided that they are compared to other cases (Herriot and Firestone, 1983). Case study approaches are common in agribusiness studies.

Case studies have "local relevance" but may not have "global relevance", be applicable to other cases, or enable generalisation. Case studies therefore contrast with experimental and large sample surveys that allows for wider inference, albeit at the cost of being difficult to apply practically to local situations and involving practical limits on the scope of questions or amount of information that can be obtained. Glaser (1994) advocates that researchers choose case studies that can be learnt from rather than whether they are representative. However, the capacity of a case to be generalised depends largely on the establishment of representativeness.

In the central planning era cattle were treated as a means of production especially for draught and transport, and were not traded (live) for meat purposes. While beef was classed as a staple product for ethnic minority areas in the central planning era, it is distinct from grains and pork which were classed as staples for the country as a whole. The cattle and beef industry therefore only recently underwent large scale development and commercialisation with the introduction of household and marketing reforms in the 1980s. In this regard, the beef industry shares similarities with a large number of other non staple, cash commodities that have proliferated and grown in relative importance in the post reform era and are undergoing specialisation and segmentation (see Chapter 4).

Beef is also a highly perishable product (like other meats and dairy) and therefore differs by degree to most crops and absolutely to fibres. Compared to most other livestock (dairy, poultry), the industry is relatively localised, although this is changing with the development of cold storage and transport infrastructure that allows for beef and cattle to be moved longer distances. Chickens and pigs are produced, processed and marketed in more intensive and mechanised systems where scale and technical efficiencies are more critical. The bovine industry is regarded in China as one of the most fragmented, small scale and "backward" livestock industries in China. There are also large differences in the structures of different livestock industries for example in the type of "dragon head" organisations prevalent in the industry.[30]

The different nature of the commodities and industries even with in the livestock sector makes it difficult and potentially dangerous to apply broad findings and responses across whole sectors. The certainty that there will be differences between industry cases is not, however, an argument against the industry case study approach. To the contrary, recognition of the differences between industries is important in developing targeted and effective findings, recommendations, policies and development projects. This is particularly important as policy makers in China—as well as development agencies—often use industries and industry policies as a development tool, with perceived advantages for industry participants (see Waldron *et al.*, 2003, Chapter 9). As highlighted in the literature review, there are relatively few in-depth rigorous Chinese industry case studies to draw on and to compare with. Thus, the conduct and comparison of more industry case studies may be a worthwhile area of research activity.

Several measures are taken in the study to facilitate comparison and contrasting between the beef and other industries. First, the literature review in the previous chapter aimed to provide a sector wide overview of the literature to provide context for this more narrow industry study. Second, the discussion in each chapter begins within a general perspective of the Chinese agricultural or livestock sector, before focusing on the case study industry. Third, comparisons are made where possible with other livestock industries that have been studied in detail (especially wool, sheep meat, dairy). Both similar and opposing findings are noted, with a view to explore explanatory causes in the interests of generality.

---

[30] In the dairy industry, relatively concentrated processors linked to other smaller milking station companies are the main form of dragon heads enterprises. In the beef and pork industry, the main form of dragon head enterprise are abattoirs. In the sheep industry, dragon head enterprises tend to be breeding companies and restaurant franchise chains, the latter sometimes integrated back to slaughter stage. In the poultry industry—and to a lesser extent the pig industry—large breeding operations also sell out young animals (hatchlings or piglets) on a large commercial basis. Textile mills and scours have traditionally been used to develop local wool production, but after their mass closure in western China, wool marketing companies are now regarded as the main dragon heads in the wool industry.

In sum, this industry case study might be useful as a point of comparison with other studies, the compilation of which can be useful to understanding of the agricultural and rural sectors. In this regard, the study aims to contribute to understanding, knowledge and theory.

3.1.3 Value of the study as an applied beef specific study

In addition to the value of the using the beef industry as a case that provides insights into the broader agricultural and rural development process in China, and as a basis for comparison with other agricultural industries, it argued that the study has value as an applied study with a focus on the beef industry itself. In this regard, another aim of the study is to draw recommendations and policy implications that are relevant to officials and economic agents in the beef industry in particular. In this regard, any improvements in understanding that can elicit improvements in policies can have a significant impact. As discussed in Chapters 4 and 5, the Chinese beef industry is growing rapidly in absolute terms and even in relative terms with the burgeoning livestock sector. Estimates based on incomplete information on the number of participants involved in the beef industry are provided in Chapter 6 to 9. Amongst the participants in the cattle production sector are unspecialised households (13.6 million), specialised households (420,000) and larger feedlots (10,000). A large number of dealers and agents are specialised in cattle trading (up to 50,000), beef and offal trading (up to 200,000) and hides. There are large numbers of people working in the slaughter sector, either in their own small slaughter operations (40,000 households) or for wage labour (150,000 people). Details of how these figures were derived are provided in relevant chapters (Chapters 6 to 9) of the book.

In addition, the impact of the study on the industry itself can be significant because of the proactive role of policy toward the industry. Like many other industries in China, the beef industry has been targeted for development under a range of direct programs (such as Straw for Ruminants as well as Favoured Areas programs), indirect programs (agro-industrialisation, agricultural mechanisation) and public programs (such as food safety). Given the policy settings, there are potentially significant benefits in a more refined and targeted approach.

3.1.4 Scope of the industry study

As implied above, a major advantage of the industry approach is that allows a micro-level analytical perspective on a wide range of issues in the Chinese agricultural and rural economy. The industry approach also has advantages in that it imposes boundaries and analytical focus in the analysis of these issues. Despite the focus, it is not possible to investigate in-depth within the scope of the book all aspects of the vast terrain covered by a rural industry the size of the beef industry. Thus while the author has investigated many of the aspects listed below, the following areas of analysis are not reported widely in the remainder of the book:

- Relatively little attention is given to beef marketing, despite its importance to the industry. Marketing aspects of the study are confined to cattle marketing (Chapter 8) and a brief discussion of beef retailing (Chapter 6). Beef marketing in rural markets is analysed in detail in Longworth *et al*, (2001, Chapter 12), while higher value beef marketing is analysed in detail in Waldron, Brown and Zhang (2007).
- By-product processing (leather, offal, medicinal products) are not addressed in any detail in this study but are discussed in Longworth *et al.* (2001, Chapter 10).
- Processed beef and foods that constitute about 10% of meat in China (State Council, 2003), are not specifically covered.
- The analysis does not include the international trade sector. Despite the size of China's cattle and beef industry in international terms (with the largest bovine herd and the third largest beef output in the world), trade represents an insignificant part of the Chinese industry. Trade aspects of the industry have been addressed in detail in Longworth *et al.* (2001, Chapter 16) and Waldron, Brown and Zhang (2007).
- Analysis of inputs and services from the extension system are not reported on in this book but are in Longworth *et al.* (2001, Chapter 3) and Waldron *et al.* (2003, Chapters 9 and 13) and in Waldron, Brown and Zhang (2007).

The study does not place strong emphasis on regional and spatial dimensions of the beef industry, which are examined in-depth in Brown *et al.* (2002b). However, Chapter 5 of this book provides a brief and updated regional overview of some of the regional issues from a macro perspective. Most of the study is focused in intensive cattle production areas in the Central Plains and Northeastern China based on agricultural systems, which have relatively developed cattle and beef industries compared to other parts of China. As a result, the study is concerned primarily with increasing efficiency, commercialisation, and specialisation of the industry. It is less concerned with the welfare value of livestock in semi-subsistent systems, which are more prevalent in the poorer and more remote parts of Southwest and Northwest China. However, the industry development paths taken by agricultural areas of the Central Plains and Northeast are likely to be replicated in the remote areas of Northwest and Southwest China, including in semi-pastoral areas.

The study is also not directly concerned with regional development implications of the development of the beef industry. That is, of the "three rurals problem" (of agricultural, rural income and rural area development), attention is focused on the former two "rurals". Analysis of regional development would require an additional analysis on the regional comparative advantage of various industry activities that are beyond the scope of analysis in this study.

3.2 A systems approach

While the industry approach was argued to be useful in addressing the research questions of this study, the approach is only useful if the industry is analysed in an ordered and disaggregated way. This is done by treating the industry as a system comprised of a series of sub-systems. The sub-systems and the relationships between them within the broader system are analysed through a ranges of analytical techniques to form a multi-disciplinary integrative study.

3.2.1 The industry system and framework

While a systems approach has not been widely applied to industry studies in China, it would appear to be well suited. Industries exhibit many of the properties that characterise a system, as adapted from Skyttner (2006).

- Systems form a dynamic and complex whole, interacting as a structured functional unit
- Complex wholes are made up of smaller subsystems
- Specialised units in the system perform specialised functions, as differentiation occurs in the system
- Energy, material and information flow between the different units and sub-systems that compose the system
- In an open system, energy, material and information flow from and to the surrounding environment via semi-permeable membranes or boundaries
- The transformation of inputs into outputs is the process by which the goals are obtained
- Regulation, or a method of feedback, is necessary for the system to operate predictably

- Systems are also often composed of entities seeking equilibrium but can exhibit oscillating, chaotic, or exponential behavior

An industry system can be conceived in a range of ways. Most systems approaches in agriculture contain biophysical components. In the case of a beef cattle system, biophysical components include breeding, feeding and disease sub-systems. Agribusiness studies usually depict supply chains in a linear way or as webs with flows of product, information and capital flowing backward and forward between various stages of production, product transformation, logistics, and supply chain actors. This agribusiness dimension is also integrated into the framework used in this study down the vertical axis of Figure 1.2 (above), which is extrapolated in Figure 3.1 (below). However, this traditional agribusiness type approach has been supplemented in this study with a focus on differentiated value or agrifood chains. As discussed in Chapters 2 (in the context of "high value agriculture") and Chapter 4, market and agrifood differentiation is an increasingly important in the development of Chinese agriculture, with far reaching implications for industry participants.

The other additional dimension of the study framework is that it incorporates the drivers and constraints to change within the industry and industry sectors. This includes drivers from both within the system (such as consumption, competition, efficiencies) and outside the industry system (such as policy, institutional and social drivers). The modernisation of agricultural industries is defined by the movement along (from left to right) of these agrifood chains. However, these are subject to various constraints including the viability of economic agents under market settings, and (land, capital, labour and human) resource constraints.

It is important to note that the framework is intended to act as a conceptual framework. As with all frameworks and theories, the framework represents a synthesis of reality in several areas. The first form of synthesis in the framework is that it differentiates between only three—low, mid and high value—agrifood chains. In reality, many more value segments can be identified.[31] In addition, the value chains are depicted in Figure 1.2 as discrete chains, where product flows vertically up through each chain. In reality, value chains form a continuum with strong relationships between the chains. For example, large modern abattoirs are established in China to service high value beef markets, but also sell mid and low value cuts into mid and low value beef markets. In the same way, modern abattoirs buy a lot of their cattle from feedlots, but also from specialised households.

---

[31] Waldron *et al.* (2003, Chapter 4) identify and map five (low, low-mid, mid, mid-high, and high) value chains.

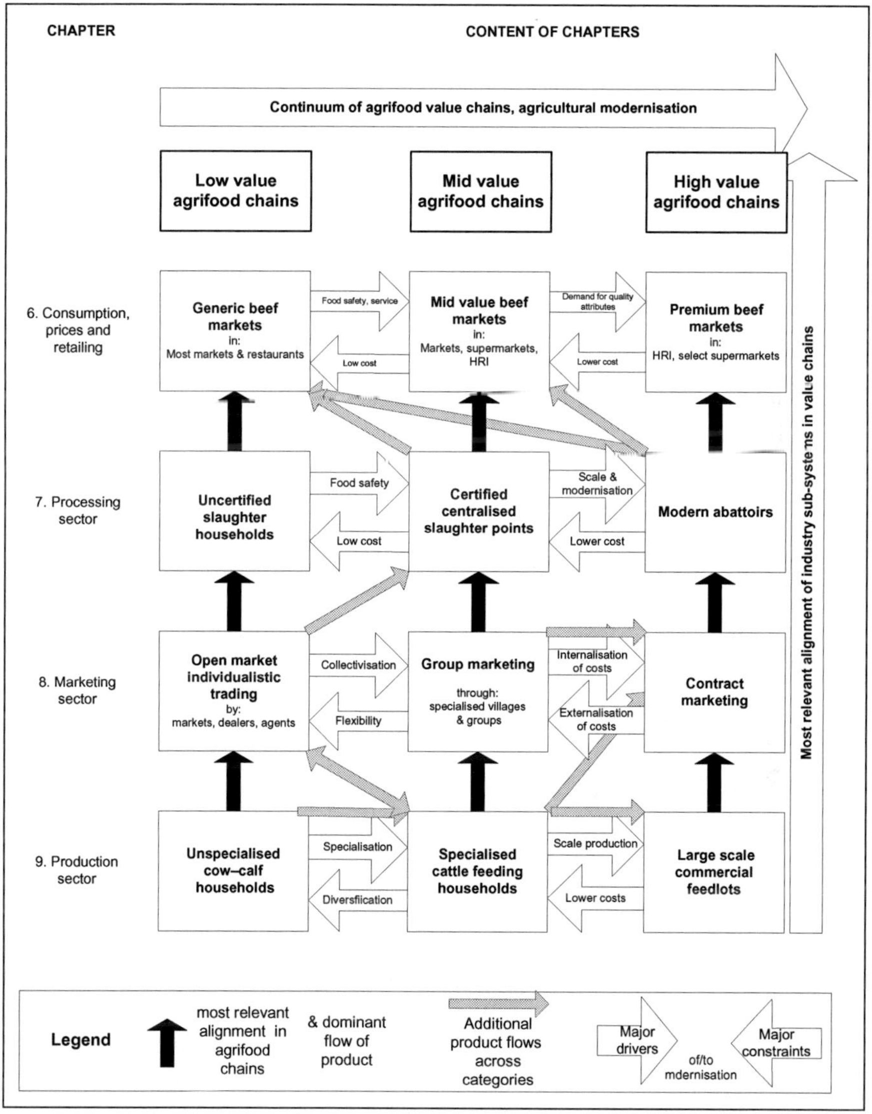

Figure 3.1: Detailed framework of the beef industry system

These cross-chain relationships are depicted in the oblique arrows in Figure 3.1. For these reasons, and due to lack of data, it is not possible to precisely or definitively quantify the size or relative importance of the different value agrifood chains, although estimates of market share are provided in Table 6.1.

The second form of synthesis in the framework is that distinctions are made between the sub-systems on a horizontal level (within industry sectors). Typologies to distinguish between different sub-systems are based on those commonly used in China between, for example, unspecialised and specialised households, or certified and uncertified slaughter units. Again, in reality these horizontal dimensions also form a continuum of economic agents. In any typology, there is likely to be blurred or overlapping boundaries, although attempts are made in relevant sections to define and differentiate between categories.

In summary, the framework itself as depicted in Figures 1.2 and 3.1 is only a representation of reality designed to facilitate and order analysis in a systematic way. However the framework is used to guide analysis within the study in a way that allows for grounded and empirical observations.

3.2.2 Analysis of the system

The treatment of the industry as a system involves the incorporation of "systems thinking" which is "a way of looking at organisations that emphasises the interconnections between parts of an organisation and external environments". It is based on the premise that the whole can not be explained by the addition of all the parts alone and that the system as a whole determines in an important way how the parts behave. This holistic or macroscopic research approach contrasts to a reductionist approach, which maintains that a system can be explained by reduction to its fundamental parts and where questions can therefore be addressed in isolation (Bertalanffy, 1976).

The analysis of the industry as a system has a number of advantages. First, it forces a disaggregated perspective that contrasts with an aggregated macro perspective from "outside" the system through macro statistics for example. Second, it allows for examination of the relationships between the sub-systems to examine how one part of the industry impacts on another. For example, policy toward modern abattoirs impacts on household cattle producers in ways discussed in Section 8.3.3. Third, establishing these relationships is useful in identifying problems and forwarding policy recommendations. Systems thinking has been argued to be "a method for solving organisational problems and helping organisations change" (Senge *et al.*, 1994).

Systems thinking provides for a large degree of flexibility in the disciplinary approach and analytical tools adopted. Almost by definition, systems are impacted by a wide range of factors, including biophysical, social, policy and market factors. The multi-disciplinary landscape traversed by an industry and the wide range of data needed to understand this landscape also limits the usefulness of a single disciplinary perspective. Disciplines that are drawn on in this study include mainstream agricultural economics, neoclassical economics (including farm budgeting), agribusiness and supply chain analysis, marketing economics, institutional economics and development economics. The analysis is also based on a technical competence in meat and livestock production and marketing.

There are strong complementarities between the disciplinary tools used. For example, the analysis of supply chains is usually associated with the field of agribusiness, but can also benefit from on-farm budgeting drawn from a long tradition of farm economics.[32] Together, these approaches provide insights into industry modernisation and implications for industry participants, including rural households. Causal relationships such as these are increasingly being discussed in the literature (see Chapter 2) and by Chinese policy makers (Chapter 4). Similarly, policy toward food safety is a public health issue, but has implications for economic structures and incentives, and therefore for rural development. Seen in this way, the multi-disciplinary approach adopted in this study acts as strength of the study, especially as it is integrated and "disciplined" within the study framework.

The nature of the systems approach also impacts on the type of analytical tools that can be used. In large, complex systems where everything is related to everything else, establishing causal explanations through mathematic testing or quantitative methods may either not be possible, or too cumbersome to be useful. Even if causal relationships can be established, they may be of minor importance when immersed within a system of many other interconnected variables (Dick and Swepson, 1997). For this reason, systems approaches—and other "holistic" approaches[33]—tend

---

[32] The complementarities and linkages in the application of the two disciplines of agricultural economics and agribusiness is overviewed in Cook and Chaddad (2000).

[33] Evolutionary economics (which can be traced to Schumpter) might be considered a holistic theory. Evolutionary economics models are open, and consider that it is hard to predict exactly the impact of a policy measure, and tend not to be mathematical (Dopfer, 2001).

to use qualitative methods which are best able to capture the reality and complexity of the system.

Systems approaches however also acknowledge the use of quantitative methods and the complementarities between the two, which can help establish causality. This study is primarily a qualitative study, based on the interpretation and ordering of interview material. However, this qualitative approach is also supplemented with the use of quantitative methods. In particular, partial budgets and sensitivity analyses are used to examine economic incentives of various economic agents (cattle producers, slaughter units) and the relationships between them. These budget analyses are integrated in so far as outputs from one sub-system (e.g. specialised cattle producers) are used as inputs into another sub-system (e.g. mid value, certified abattoirs), and where the input–output values (e.g. product attributes and prices) are consistent. This allows for a significant degree of integrated quantitative analysis across the whole industry system, and forms a single substantive quantitative analysis of the study.

3.3 An inductive research approach

This section overviews the philosophical approach taken to the research in this study. The study is described as being inductive, positive and applied and in nature. This research approach is also reflected in the projects and the data on which the study is based as discussed in Section 3.4. The research methodology employed in this study can be seen as a continuation of that used by the broader research team in an earlier era, as outlined in Longworth and Williamson (1993, Chapter 2).

One approach to research is to first derive research questions or hypotheses, then develop the methods to address the questions or test the hypotheses, and then go about collecting the data. This "reasoning down" process from hypothesis to data is known as deductive reasoning, or logical reasoning from general principles (Creswell and Clark, 2003). In the case of this study, however, a large body of data was first collected and analysed (for other studies mentioned below). In the process of collecting and analysing the data, research questions and methods reported in this study were developed and refined. Opportunities then arose to collect and analyse more targeted data for this particular study. Several such iterations occurred, as described below. This iterative, "reasoning up" process from data to hypothesis is known as inductive reasoning (Creswell and Clark, 2003).

Inductive reasoning is aligned with several other research approaches. The first is scientific realism, where "research data has allowed for the development of theory, and where the theory is aligned with reality". The second is grounded theory which encourages the researcher to "ground" themselves in empirical data before developing a theoretical framework. The theory is built progressively as the study proceeds (Glaser, 1994).[34] Third is an action research approach, which also draws conclusions which are data based and emerge slowly over the course of the study, and where data analysis and theory building occur at the same time as data collection.[35] The fourth approach is the use of participatory research methods, which is typically led by individuals working in teams (to improve the way they address issues and solve problems) with informants involved as active participants in the research process (to provide more information about the situation). This "collaborative inquiry" involves research *with* people rather than reflective knowledge conducted *on* people. Participants are therefore involved in active moment to moment theorising, data collecting, and inquiring (Heron, 1971).

Inductive research is most useful when no one theory is able to explain the research question. In this case, multiple theories can be drawn on and developed, and "space" between the existing theories filled (Glaser, 1994). However, inductive research methods tend to be used more by practitioners than by theorists. In this regard, Glaser takes a "practical" view of the role of theory, where the adequacy of the emerging theory is judged by the criteria

---

[34] Glaser advocates a research process of simultaneously collecting data, talking notes (on the content of the interviews), coding the data (by sorting it into categories), and writing memos (on the possible links between the categories). Researchers compare and analyse the data to identify categories and trends, until categories and trends regularly appear. Over time no further categories emerge leading to "theoretical saturation". A "core category" also emerges, which is of central concern to the people in the situation, and to which many other categories are linked. As further information is collected it is compared to the emerging theory. The theory is refined to take account of the additional information.

[35] A central tenet of action research is that research is conducted in cycles or spirals. Various types of cycles are described starting from Lewin (1946), to the "experiential learning" cycle of Kolb (1984), to the iterative comparing, disconfirming and explaining of sets of data of Strauss and Corbin (1990). However, all action research approaches are based around the two way alternation of action (change) and critical reflection (analysis of what has happened and planning for the next action). A researcher can enter a research area without a precise research question or a precise methodology. As data is collected in the research area, understanding accumulates. This allows for modification of methods, data and interpretation. Regular critical reflection tested in action provides chances to correct errors and is the source of much of the rigour in action research.

of whether it fits (is faithful to the every day reality of the subject areas), and it works (leads to reliable predictions which are of relevance and value to those involved in the area of the study). If the theory fills these criteria, it will assist people to make sense of the situation and manage it better. This study represents an applied rather than theoretical study in so far as makes recommendations to improve a particular development strategy and policy settings. However, in doing so, implications for theory arise, as outlined in Chapters 2 and 10. It is possible that the holistic, inductive and grounded nature of the research in this study can add to understanding especially to that from more reductionist, deductive and reflective studies.

Another classical distinction in research approaches is that between positive research (what is) and normative research (what ought to be). Positive economics describes and explains economic phenomena, focuses on facts and cause and effect relationships and includes the development and testing of economic theories in a value free way. Friedman (1953) argues that economics as a science should be free of normative judgments if it is to be objective and is to inform normative economics. Unlike positive economics, normative economics incorporates value judgments about what the economy should be like or what particular policy actions should be recommended to achieve a desirable goal. Normative economics looks at the desirability of certain aspects of the economy and underlies expressions of support for particular economic policies.

The analysis in the body of this study aims to be positive in nature. In particular, Chapters 4 and 5 describe trends in the Chinese agricultural sector and the beef industry based on official policy documents and statistics. Chapters 6 to 9 analyse industry agents and systems *in situ*, including the economic, social and policy environment in which they operate. The normative aspects of the study arise most obviously in the final recommendations and policy implications chapter (Chapter 10) where the study prescribes particular strategies, policies and institutional measures that are aligned not only to market forces but also to the interests of industry participants from rural areas. However, the broader study may be regarded as having a normative element in so far as it is based on the premise that policy toward the industry should take into account the participation and impacts on rural participants.

3.4 Data collection and analysis

The broad research approach described above is reflected in the research projects and the data collection methods used in the study described in this section.

3.4.1 Research projects drawn on for the study

The broad research approach described above for the study has roots in a longer term research program as summarised in Table 3.1. The author of this study has been closely involved in various projects on the Chinese beef industry, each of which had a different focus including: a socio-economic focus (1997–99); an agribusiness focus (1997–98); an impact assessment of the socio-economic project (2004); and an update on the previous agribusiness project (2006–07). The different foci are reflected in the diversity of major research outputs, on development aspects of the industry (Waldron *et al.*, 2003a); agribusiness aspects (Longworth *et al.*, 2001; Waldron, Brown and Zhang, 2007); and regional aspects (Brown *et al.*, 2002b).

While each of these major projects and publications had a different focus, they have effectively been treated as an integrated whole. The author has been engaged full time in all projects, fieldwork for the projects was conducted simultaneously and there was a seamless interchange of data between the projects. The multi-disciplinary nature of the study was also forged by the broad range of interviewees interviewed on the projects, across diverse areas and over a long time period. This allowed for observation of the full range of factors that impacted on the industry. The multi-disciplinary and integrative nature of the projects, especially the interplay between socio-economic (developmental) and agribusiness aspects, is continued in this study.

The enormous amount of data collected on those projects means that the data has been heavily synthesised to be presentable within the confines of this study. Little of the data collected in the projects outlined below—especially from fieldwork—is actually presented in this study. However, references are made to previous work of the author, where more direct use of the data was made.

While the study draws heavily on this previous work of the author, it also aims to extend that work and differs significantly in several ways. First, the study presents a different analytical framework and perspective to that used in other project publications. Second, the study draws on a different body of literature. Much of the literature overviewed in Chapter 2 was not prominent when the original beef projects began in the late 1990s. These studies have helped frame the research questions and approach, and provided different theoretical perspectives and "theoretical gaps" to fill. Third, the study also represents an update of the previous published work by the author. In particular projects done in 2004 and 2006 have allowed for the incorporation of new data and allowed for observation of medium term (10 year) development trends in the industry. This has also raised new questions and generated new findings. In this regard, this study draws heavily on information collected for the report by Waldron, Brown and Zhang (2007). Fourth, and very importantly, compared to previous beef publications by the author, this study provides more comparative and contextual information gained from projects on other ruminant livestock industries in which the author has been involved, including those on the wool industry (Brown, Waldron and Longworth, 2005), sheep meat (Waldron, Brown, Longworth *et al.*, 2007) and grasslands (Brown *et al.*, 2008). Besides enabling ongoing contact with the beef industry, these studies also enabled comparison across industries, some of which have been incorporated in this study.

3.4.2 Data drawn on for the study

When the first round of beef projects were initiated and conducted in the late 1990s there had been very little previous systematic investigation of the still emerging Chinese beef industry, especially from an economic perspective. Other studies were largely confined to technical aspects of cattle and beef production and made heavy use of macro statistics. The projects therefore started on a base of inaccurate and highly aggregated statistics and incomplete anecdotal information. One exception was Lu (1994) which acted as the genesis for the beef projects on which this study is based. In general, however, an understanding of the issues confronting the beef industry had to be built "from the ground up" in an iterative process in order to build knowledge and propose recommendations for the industry. Indeed, these tasks were the objectives of the initial projects.

The type of data and information collection processes by which this knowledge of the industry was built is overviewed below. It is a described as a triangulation of three types of data namely: macro-level statistics and policy documents; micro-level budget data; and interview material.

Statistical data and policy documents

Much of the design of the early beef projects that started in late 1990s were based on macro statistics reported in various yearbooks, including the China Statistical Yearbook, the China Agricultural Yearbook, the China Animal Husbandry Yearbook, China Rural Statistics Yearbook and provincial and city statistical yearbooks. Analysis of statistics was done by the project team as a desktop activity early in the project, assisted by Chinese collaborators that were very familiar with the data. Importantly, collaborators were also able to calibrate the data with a strong practical understanding of the livestock sector, including some understanding of the beef industry. The macro statistics provided an important starting point in providing context, identifying trends for closer analysis and choosing fieldwork areas.

Official statistics remain the main source of macro-level information with adequate spatial and longitudinal coverage to identify broad macro trends across entire industries. Carefully interpreted, this can provide useful context for other forms of more disaggregated data. However, as discussed in more detail in Chapter 5, official Chinese statistics for agriculture including those for livestock—livestock production, turnoff, livestock product production and consumption—have severe limitations. With regards to accuracy, growth rates in production statistics were and remain overstated and, of most concern, the degree of overstatement varies over time and region. While this was suspected at the beginning of the projects, the extent of the overstatement in the late 1990s was unprecedented even to seasoned Animal Husbandry Bureau officials. Fuelled by the fervour to meet aggressive industry plans, and to report favourably on strong industry policies, production statistics were overstated for beef by up to 30%, a magnitude higher than other livestock industries. The extent of the overstatement was highest in intensive beef production areas, many of which chosen as fieldwork in the initial years areas, including Shandong and the infamous Fuyang Prefecture in Anhui Province.

Fortunately the China Agricultural Census was conducted in the late 1990s and the statistics were revised downwards by up to 30%. After this downward revision, the statistics went through another decade of overstatement. Aware of this, Animal Husbandry Bureau officials worked on the assumption that production statistics for cattle and beef—and other livestock—were about 20% overstated. This prediction proved was reflected in the Second National Agricultural Census when official Chinese statistics for the years 2000 to 2007 were revised. As was the case in the First Agricultural Census, statistics for the beef industry were revised down dramatically. For example, in 2006 bovine numbers were revised down by 25% and beef output by 23% (China Animal Husbandry Yearbook, 2009).

In addition to being inaccurate, the official statistics are limited by their levels of aggregation. For example, prices reported in the China Animal Husbandry Yearbook are for average mixed grade product only and do not report on price-grade differentials. In addition, some data series have changed to an extent that precludes longitudinal analysis. For example, scale of beef cattle production data was added to the China Animal Husbandry Yearbook in the early 2000s, but in an effort to make the data more relevant to the fast changing industry, the scale categories and units of measurement (i.e. stock numbers versus turnoff) have changed several times.

Given the limitations of the official statistics, various industry indicators were collected by the project team in all fieldwork areas. In 1997 to 1999, regional "data only" surveys were conducted from province to township levels, covering a period of ten years. Some of the data is presented and analysed in Longworth *et al.* (2001) and Brown *et al.* (2002b).

Policy documents (including plans, edicts, laws, policies, statements and "opinions") relevant to the industry were routinely collected for all beef projects. Interviewees were asked before and during fieldwork for policy documents, some of which were publicly available and some of which were "internal" and not therefore able to be quoted or cited. Policy documents were able to be compared over a large number of areas, administrative levels and organisations. This proved useful in incorporating an "official" perspective of the industry and placing economic analysis in context with the broader policy environment. Some of these documents are cited in Chapters 4 to 9, though due to space constraints citations are restricted to central level policy documents.

Like official statistics, policy documents provide a guide to current and future developments in the industry, but need to be interpreted carefully. For example, plans should not be interpreted literally but for their insights into the "official line" and objectives. Funding levels indicated in documents are usually planned rather than actual. Funding often appears to be substantial but the actual impact can be limited when distributed over an enormous number of administrative, service and economic agents. Policies expressed publicly or at high levels can be virtually unrecognisable at local levels, creating what Longworth and Williamson (1993) describe as a "policy mirage".

Surveys and budgets

Several surveys of economic agents were conducted in related research projects and drawn upon in this study. Surveys were conducted on various types of cattle producers, including unspecialised households, specialised households and feedlots in agricultural areas. These surveys were designed after understanding of the issues had been gained in the initial year of fieldwork in 1997. About 25 units in each category were surveyed in 1998 and 1999, and the information and data was collected to construct "representative budgets" of each type of unit. Synthetic budgets of the operations of the representative units were generated and used to further understand the economic incentives facing typical unspecialised cow–calf households, specialised fattening households and feedlots. The results of the systems and the budget analyses were written up in detail in Longworth *et al.* (2001, Chapters 5, 6 and 7). The budgets were then updated from information gained on fieldwork in 2006 and run again to produce the results presented in this study. The details of the budgets are provided in Boxes 1, 4 and 5.

The small sample size of the surveys precluded a statistically valid study but formed a set of numbers to be verified through other sources. First, the questionnaires were enumerated by the author himself (rather than by collaborators) as one part of interviews that lasted on average half a day. The broader open-ended discussion in the interviews allowed for an in-depth understanding of production systems and the environment that supplemented the numerical data. Second, advice on budget parameters was inputted from other industry stakeholders, such as livestock extension staff, officials and researchers. Third, the budgets drew on results from various feed trials and technical data. This combined data collection process generated "synthetic" or "representative" budgets. It is argued that the supplementary input from various industry stakeholders overcame some of the problems associated with the small surveys sample and reinforced the validity of the data and the understanding of the systems.

A second type of budget analysis conducted was on various types of slaughter units. Abattoir budget parameters and values were derived not from formal surveys but constructed through interviews with more than 40 slaughter units. These "synthetic" budgets are presented in detail in Longworth *et al.* (2001, Chapter 9). Parameters were updated on fieldwork in 2006. Third, beef consumption surveys were also conducted in northern Jiangsu Province, outside the major cosmopolitan consumption cities like Beijing and Shanghai, and reported in Longworth *et al.* (2001, Chapters 13 and 14). Unlike the collection of the other survey and information drawn upon in this study, the author was only indirectly involved in these consumption surveys.

Budget analysis was chosen as the main form of quantitative analysis used to complement the mainly qualitative analysis in the study for several reasons. First, budgeting enables a focus on the systems and incentives of specific economic agents. Second, there is little rigorous budgeting done in the beef industry in China. Other studies by, for example Wang and Zhang (1994) and World Bank (1999), were simplistic. Third, industry participants (including policy makers) relate easily to budgeting exercises and "bottom line" indicators and therefore the budgeting approach facilitated communication with the industry and informants. Fourth, as synthetic budgets were constructed for several categories of economic agents that span sectors and value chains, input–output parameters were coordinated to synthesise the broader industry supply chain.

Fieldwork and interviews

The major method of building knowledge of the industry in the projects was through fieldwork, especially interviews with industry stakeholders. The emphasis on interview data is only partly because of the lack of, or limitations in, secondary data and statistics. Interview material allows participants to construct responses which are meaningful and valid to them within their social context (Williams and May, 1996). Industry stakeholders interviewed for the study, especially at local levels, were almost always highly knowledgeable about their local environments. Their pragmatic, detailed and empirical knowledge provided important insights that can not always be captured through broader or formal survey-based studies.

Fieldwork for the four beef projects were conducted in 18 provinces and regions of China. Fieldwork covered all five "beef zones" in China across highly variable systems and levels of development. Within each of these zones and provinces, fieldwork was conducted in the most intensive beef prefectures and counties. However, even the most developed beef counties in some provinces (such as Yunnan and Sichuan) were highly undeveloped. Furthermore, fieldwork conducted for other (non beef) projects allowed for observation of the beef industry as an undeveloped industry of secondary or negligible importance in the local livestock economy.

Fieldwork "telescoped down" sequentially from central level to provincial, prefecture/city, county and township to village levels, with fieldwork at all levels. This allowed for in-depth study of the institutions themselves and their horizontal and vertical relationships. In total, about 500 interviews were conducted for the beef projects alone spanning more than a decade, and many more for other related projects. Within the broader research team, the author was involved in virtually all this fieldwork, making a cumulative period of more than one year in the field, allowing for immersion in and understanding of many of the nuances of the beef industry.

All fieldwork was hosted by the Animal Husbandry Bureau at various levels. As such, this institution acted as the primary source of information, organisation and logistics. This also has implications for the study. In particular, the Animal Husbandry Bureau selectively provided information and access to other interviewees, including other government departments and households. However, the Animal Husbandry was just one of a large number of organisations interviewed. Interviewees typically included various (administrative and service) units in the Animal Husbandry Bureau, local government or Party officials, the Economic Commission, Industry and Commerce Administration Bureau, Environmental Protection Bureau, Poverty Alleviation Bureau, Supply and Marketing Cooperatives, General Food Companies, local banks and credit cooperatives, breed stations, various industry associations, academies, abattoirs, food and by-product processors, retail markets, supermarkets, live cattle markets, feedlots, household producers. While there were some regional variations, interviewees were similar across regions, allowing for comparison across regions. Access to this diverse range of interviewees was only possible through involvement in large collaborative research projects.

In the first year of the projects, interview questionnaires were developed for the specific organisations and administrative levels interviewed. The questionnaires were designed to act as checklists and prompts in semi-structured interviews that usually lasted for half a day. Large scope was provided to interviewees to express opinions and pursue themes, especially at the beginning for the interviews. The growing knowledge base throughout the project allowed for the refinement of interview questionnaires. By the third year, questions became more targeted to confirm or negate questions, to fill knowledge gaps for planned research outputs and to pursue local variations of themes that arose in other areas.[36]

Interviews were in many cases linked with field visits. The opportunity to observe interviews and economic agents *in situ* provided important detail, context and "reality checks", which are necessary in the development of case study material. At the end of fieldwork visits to particular areas (to, say, a county for a week), hosts usually asked the research team for their opinions

[36] Open-ended interviews in China can have limitations because interviewees tend to talk in generic terms or from documents prepared for high level (non expert) officials. Interviewees will often not volunteer information, or know what information to volunteer, to foreigners. However, most interviewees willingly attempt to answer specific questions, especially if they are well founded. Interviewees also try to ascertain the level of expertise and knowledge and the areas of interest of foreign interviewers before responding, or whether the visitors were "industry/discipline outsiders" (*waihang*) or "industry/discipline insiders" (*neihang*).

on the local industry, or for comparative information on the Australian and international industry. These exchanges often became interactive informal workshops between local hosts, Chinese collaborators and the Australian research team members. Importantly, it also allowed for some of the research team to "test" and refine assumptions and recommendations at a local level. These relationships were further developed through larger formal national level industry workshops and conferences.

In addition, return visits were made to five areas selected as fieldwork sites for various beef projects, some of which (Shandong and Henan) were conducted eight to nine years apart. The repeat visits allowed repeat interviews with the same organisations, and therefore direct comparison over time. In those repeat visits, the field reports from previous visits were consulted to frame questions. In this regard, the methods used for the research are closely aligned to action research methods overviewed above.

Australian members of the team were present at all interviews. Interviews were conducted in Chinese, with transcription into English by the author, and cross checked for accuracy by colleagues and collaborators, who also added comments and interpretations into the transcripts in a form distinguishable from the original interview text, and were often subsequently discussed. In this regard, and many other ways, working in a team was invaluable and of great benefit to the research. Other information collected on fieldwork (such as data, policy documents, brochures and photos) was attached to the field reports. Transcripts were compared across organisations and regions. Field reports were then sorted according to keywords and entered into a "fieldwork database". Numerical information was extracted and entered into a "margins database". Through this process, field reports allowed for the identification of patterns and themes. Publications were often drafted through the compilation of field reports and analysis refined by searching for consistencies and differences in the reports.

Validity and reliability of fieldwork data

It is widely recognised that visitors (government, business or research) are regularly given a distorted or incomplete picture by local hosts in China. The primary reliance on fieldwork information in this study therefore raises questions about potential data error or bias. The research team were conscious that many interviewees portray an overly positive picture of the industry in order to maintain "consistency" with the overstated statistics and upbeat reports sent to higher level officials. The presence of senior central level Chinese researchers in the research team also provided local areas with another line of communication to central level officials and an opportunity to lobby for projects and funding which were generous in the beef industry especially in the 1990s. Local officials were also sometimes interested in hosting the foreign research team on the (mistaken) belief that it would help attract foreign investment, especially in the processing sector. As a result, local areas tended to exaggerate data on items like cattle availability and productivity, and the organisational capacity for government to ensure supply for any such investment in the processing sector. The overstatement of local raw material supply by local officials has proved a serious problem for many investments in the agricultural sector of China.

While much of the information provided was overtly positive, hosts also tended to strategically identify problems that they thought the research team could communicate to higher level officials. Many statements went along the lines of "even though we have developed the industry rapidly, we have a lot more to do". Most of the problems identified were related to technical and infrastructure aspects of the industry (especially breeding and services) partly because this provided opportunity for tangible resourcing from which local government would benefit. Local government almost always identified a lack of modern processing capacity and value adding activities as a constraint to industry development. This was at least partly motivated by incentives of local officials to participate in investment activity, and because enterprises can be taxed locally and provide a vehicle for claiming personal expenses. Such incentives biased the information provided and the identification of problems. For example, officials were rarely interested in scaling down production, or activities that were not lucrative for local government (such as low value processing), even if those activities seemed appropriate.

In addition, some local areas tried to actively conceal information. For example, it was difficult to visit slaughter households because they are a legal grey area, and because of the "impression" it would leave on the foreign guests about hygiene conditions in China. In the lead up to China's accession to the World Trade Organisation, there was a reluctance to discuss distortionary policies (such as underwriting of loans by government for agricultural enterprises). It is also pertinent to note that the province or prefecture Animal Husbandry Bureau that hosted the fieldwork visit dispatched an official to accompany the research team on fieldwork, often over a two to three week period. The role of this official ranged from managing logistics to acting as "minders" to conceal information, but in many cases these people became willing and friendly providers of invaluable local knowledge.

While these problems are ubiquitous in China, they are most pronounced and likely to distort research on short, one off visits. The sheer volume of fieldwork conducted for projects drawn on in this study decreased the prevalence of the biases and enabled the problems to be managed in a number of ways. Experience enabled the identification of information that was likely to be biased, such as those mentioned above. Other likely areas of bias or misinformation were identified through comparison of fieldwork information across regions, down through administrative levels (for example prefecture to county) and over time (especially in repeat visits). Inconsistent information in field reports often became the stimulus for further questioning when subsequent opportunities arose either through return visits or from other sources.

The large number and diversity of interviewees, the length of time spent in fieldwork areas and the number of first hand and visual observations made on field visits, made it almost impossible for local authorities to consistently misreport and conceal information. Long visits also allowed the identification of organisations and people that were most likely to engage in frank discussion. Fieldwork provided multiple opportunities (in meetings, banquets, cars, field visits) for informal discussion. Many industry stakeholders such as technicians (in veterinary or breed stations for example), business people (in feedlots and food companies for example) or individuals (in markets, households) do not know or are not interested in the "Party line" and tend to respond naturally to the right questions. This has also increasingly been the case for officials over the last 10 years, especially in more developed parts of China. This may be because officials are less interested in external funding and investment or because they are becoming more professional and more interested in external opinions and genuine exchange. In addition, it was explained to interviewees that fieldwork visits for projects funded by the Australian Centre for International Agricultural Research were research projects supported by a government to government bilateral protocol and signed off by relevant ministries. Interviewees responded differently to these visits than they do to visits for development or commercial projects, especially in terms of engaging about industry and development problems.

Another important aspect is the credentials and contacts of the research team. In this regard, Chinese collaborators in the research team played a critical role. Chinese collaborators often had pre-existing contact or relationships with Animal Husbandry Bureaus (especially at province or prefecture level) or were known experts in the Chinese livestock sector. Local hosts were therefore likely to engage in frank and advanced exchange.

Information known by the Chinese collaborators were often communicated to the Australian team during or after formal interviews. Except on return visits, the foreign members of the research team were unknown to local hosts. The foreign members of the research team were introduced as "foreign experts" although this was a rhetorical title until proven differently.

The capacity of the author to communicate directly and fluently in Chinese, including the use of Chinese livestock and administrative terminology, benefited the research. Fieldwork exposes researchers to a new and complex mass of verbal, written and other sensory messages. Language skills and cultural familiarity increases the capacity of the researcher to receive and interpret these messages, so as to better understand the research environment, increase knowledge and do better research. Association with cultural and language settings also assist in the development of trust and relationships, in "getting things done" and in phrasing questions and dialogue in a way to which interviewees can relate and respond more openly. Direct dialogue in Chinese can also increase the momentum of dialogue, save translation time and reduce translation errors.

Inevitably, however, foreigners are viewed as such and will misinterpret events and language, especially in remote rural areas with localised dialects, accents and cultural practices. Some of the Chinese collaborators working with the Australian researchers were outsiders to fieldwork areas and were in some cases also subject to the same limitations. However, most of the senior Chinese project partners involved in the beef projects were very experienced in the Chinese "livestock fieldwork environment" and were able to fill gaps of understanding of the author and colleagues, which grew smaller over time.

Cross-verification of data

It was pointed out that this study uses three different broad categories of data: macro-level statistics and policy documents; micro-level budget data; and interview material. It is argued that measures were taken to address the limits and increase the validity of each type of data. However, this section argues that the process of cross-verification (or triangulation) of the three forms of data provides additional validation. Triangulation overcomes the problems of dependence upon a single type of data and analysis, which are subject to some degree to theoretical and methodological criticisms. If a research questions can be confirmed by two or more independent measurement processes, then the validity and reliability of the conclusions are increased. As stated by Webb *et al.* (2000) "The most persuasive evidence comes through a triangulation of measurement processes. If a proposition can survive the onslaught of a series of imperfect measures, with all their irrelevant error, confidence should be placed in it. Of course, this confidence is increased by minimising error in each instrument and by reasonable belief in the different and divergent effects of the sources of error".

Triangulation of data is a feature of this study, including that across chapters. For example, carcass weights derived from macro data (Section 5.4.4) are consistent with cattle weights used in budget analysis (Section 9.2.2), carcass purchases of beef traders (Section 6.1) and visual observations on fieldwork, all of which confirm downward trending cattle slaughter weights and ages. The reason why this might be the case is explored in (in Section 9.1.2) where budget analysis indicates that cattle feeding in unspecialised household systems is unprofitable if opportunity costs of labour are included in the economic rationale of the households. This is consistent with macro-level scale of production level data (in Section 5.4.2) showing that small cattle raising households are leaving the industry in large numbers. The change in cattle supply and price impacts heavily on the viability of abattoirs, especially given tight margins (Section 7.3.2) and policy distortions (Section 7.3.1). The issues of downward trending numbers of cows, cow–calf raising households, the lighter turnoff weights of slaughter animals and the low returns in the abattoir sector were the main concerns of industry officials—and especially abattoirs—at the China National Beef Industry Conference held in July 2008 and were discussed in Waldron (2008) and Liu (2008), based on Waldron (2007).

Longitudinal dimensions of study

Another feature of the study on which this study is based is that all three categories of data discussed above have been collected over a 10 year period. The mid term time period is sufficient to conclude that the research has not been overly influenced by one off events such as changes in policy, price, economic, technological or climatic shocks. There has also been change in the nature of the industry from an infant industry in the 1990s to a maturing and modernising industry now. Thus, the study incorporates a strong diachronic dimension and is therefore different to the synchronic "snapshot" of many industry studies.

The advantages of longitudinal research are supported in the literature. In research situations where it is not possible to control respondents and control independent variables, a more historically oriented and more qualitative evaluation is of greater value (Weiss and Rein, 1970). Longitudinal design of the quasi experimental design group offers the greatest range of applicability to both biophysical and socio-economic investigations. A series of observations may be necessary to catch changes in behaviour over time and only after a reasonable amount of time can direct causal relationships be inferred from the data (Wall and Williams, 1970). With regard to the final point, the time period covered over the study facilitates the retrospective tracing of phenomenon in the industry to causes. That is, there are various paths that the industry did or could have traversed in its development. As suggested above, the development path taken has influenced by a wide range of factors, including policy and institutional factors. Incorporating these factors into analysis to account for the reality of the development of the industry necessarily involves a multi-disciplinary and integrative approach.

Table 3.1: Research projects, partners, fieldwork, data and research outputs related to study

<table>
<tr><th>Year</th><th>Project & agency</th><th>Partners</th><th>Fieldwork undertaken by author</th><th>Data</th><th>Selected research outputs</th></tr>
<tr><td>1997–99</td><td>Analysis of socio-economic developments in the Chinese cattle & beef industry. Australian Centre for International Agricultural Research (ACIAR)</td><td rowspan="3">• China Agricultural Economics Group
• Various Australian consultants & students
• Chinese Academy of Social Sciences (Rural Development Institute)
• Chinese Academy of Agricultural Sciences (Institute of Agricultural Economics)
• Ministry of Agriculture (International Cooperation Division)</td><td rowspan="2">• Interviews with Animal Husbandry Bureaus, Internal Trade Bureaus, Industry and Commerce Administration Bureaus, local government and leaders, General Food Companies, Supply and Marketing Cooperatives, Credit Cooperatives, Agricultural Banks, abattoirs, slaughter households, leather companies, cattle groups, cattle households, live cattle markets, beef markets, various academies etc.
• Full range of administrative levels (central, provincial, prefecture, county, township, village) in 14 provinces/regions
• 1997: 3 months in Beijing, Inner Mongolia, Anhui, Jilin, Shanghai
• 1998: 3 months in Beijing, Shandong, Hebei, Liaoning, Jiangsu, Hong Kong (plus other partner trips to Guangdong & Guangxi)
• 1999: 3 months in Beijing, Henan, Anhui, Jilin, Sichuan, Yunnan</td><td rowspan="2">• More than 400 interview transcripts
• Numerous case studies or producers, companies & organisations
• “Data only” questionnaires of fieldwork regions (all administrative levels from national down to township levels)
• Database of official statistics
• Surveys & development of representative budgets of unspecialised cow–calf households, specialised fattening households, feedlots
• Representative budget of Chinese abattoir operations
• Consumer surveys in Jiangsu</td><td rowspan="2">Liu (2000)
Zhang et al. (2000)
Longworth et al. (2001)
Brown et al. (2002b)
Brown et al. (2002a)
Longworth et al., (2003)
Waldron et al. (2003a)
Waldron et al. (2003b)</td></tr>
<tr><td>1997–98</td><td>Analysis of agribusiness developments in the Chinese cattle & beef industry. Meat & Livestock Australia (MLA)</td></tr>
<tr><td>2000</td><td>Extension of ACIAR project</td><td>2-day workshop for 75 key industry officials, researchers & business people</td><td>Exchange of information between industry leaders & projects partners</td><td>Zhang & Longworth (2000)</td></tr>
</table>

Year	Project & agency	Partners	Fieldwork undertaken by author	Data	Selected research outputs
2004	Impact assessment of: Analysis of socio-economic & agribusiness developments in the Chinese cattle & beef industry (ACIAR)		2004: 7 days in Beijing & Henan	Return visit to previous fieldwork region & organisations to report on developments in the industry	Waldron *et al.* (2004)
2006–07	Update on developments in the Chinese cattle & beef industry of relevance to the Australian industry (MLA)		2006: 2 weeks in Beijing & Shandong	• Return visit to previous fieldwork region & organisations to report on developments in the industry • Update data, policies & budgets	Waldron, Brown & Zhang (2007)
Other projects not specific but related to beef					
2001–04	Chinese Wool Textile Mills: Economic Analysis of Fibre —Input/ Textile Product Selection & New Processing Technologies (ACIAR)	Research Centre for the Rural Economy	2001: 2+ weeks in Gansu pastoral areas, plus Beijing & other industrial areas 2002: 2+ weeks in Xinjiang pastoral areas, plus Beijing & other industrial areas 2003: 2+ weeks in Inner Mongolia pastoral areas, plus Beijing & other industrial areas		Brown, Waldron & Longworth (2005) Brown, Waldron, Longworth *et al.* (2005) Zhao *et al.* (2005)

Year	Project & agency	Partners	Fieldwork undertaken by author	Data	Selected research outputs
2003–04	Analysis of agribusiness & policy developments in China's sheep meat industry of relevance to the Australian industry (MLA)	Chinese Academy of Agricultural Sciences (Institute of Agricultural Economics)	2003: 1 week Shandong, plus Beijing 2004: 1 week Inner Mongolia, plus Beijing		Waldron, Brown, Longworth *et al.* (2007)
2005–08	Sustainable development of grasslands in western China (ACIAR)	Research Centre for the Rural Economy	2005: 3 weeks Inner Mongolia, plus Beijing 2006: 3 weeks Xinjiang, plus Beijing		Brown *et al.*, (2008)

CHAPTER FOUR

MAJOR TRENDS AND POLICIES IN CHINESE AGRICULTURE AND LIVESTOCK

The aim of this chapter is to overview major trends and policies in the agricultural and livestock sectors, especially in relation to China's agricultural modernisation strategy outlined in previous chapters. It also aims to provide context for the more specific beef industry analysis presented in subsequent chapters. To this end, Section 4.1 overviews sectoral change in Chinese agriculture, while Section 4.2 overviews some of the major forces impacting on Chinese agriculture, including segmentation, modernisation and integration.

The chapter provides an "official perspective" of development of the Chinese agricultural sector in so far as it draws primarily on macro-level statistics and policy documents. Two documents that represent significant new directions in Chinese agricultural policy in recent history are particularly important, namely: the "Structural Adjustment" theme of the 1999 Rural Work Meeting; and the "Modern Agriculture" theme of the 2007 Number One Document.[37] For more detailed and localised policy analysis in the Chinese livestock sector see Waldron *et al* (2003a, Chapters 9–13), Waldron, Brown, Longworth *et al* (2007, Chapter 5), Waldron and Brown (2006) and Brown *et al.* (2008, Chapter 4).

Analysis in the chapter shows several main points. First, it shows that Chinese agriculture is becoming more diversified in its sectoral and industry composition, leading to the development of previously small and uncommercialised industries like beef. Second, from a background in centrally planned and semi-subsistent agriculture, all agricultural and livestock industries are to varying extents undergoing a process of

[37] The annual Central Rural Work Meeting of the Central Committee of the Chinese Communist Party and the associated Number One Documents (1982 to 1986 and revived in 2004 to 2007) jointly released with the State Council, represent the highest level of agricultural policy making in China. For further information on agricultural policy making in China see Zhu and Tian (1989) and Du (2006). These high level directives are implemented through an enormous number of decrees, laws, regulations, standards, plans, programs, policies and projects. These form a "policy hierarchy" as detailed in Brown *et al.* (2008, Chapter 4). Because there are too many such policies to overview, only the main policy documents are mentioned in this chapter.

diversification, segmentation, modernisation and integration. Third, these changes are not just a result of market forces, but are also forged in one way or another by policy and institutional settings. That is, policy settings can and do alter incentives for producers and other industry participants.

4.1 Changes in the sectoral and industry composition of Chinese agriculture

This section overviews the diversification of the Chinese agricultural sector due to the emergence of a plethora of previously small industries like beef. The diversification means that the agricultural sector is becoming increasingly complex, offering new development opportunities and requires new and more sophisticated policy responses.

4.1.1 Sectoral composition of Chinese agriculture

In the context of absolute growth in the agricultural sector, China's agricultural sector is undergoing rapid diversification as indicated in Figure 4.1. The diversification is underpinned by the relative decline of the cropping sector. Even within the cropping sector, non-grain products—such as fruit, vegetables and cotton—have grown much more rapidly than grains. Even within the grains sector, there has been a shift from food grains to feed grains (Zhou and Tian, 2005).

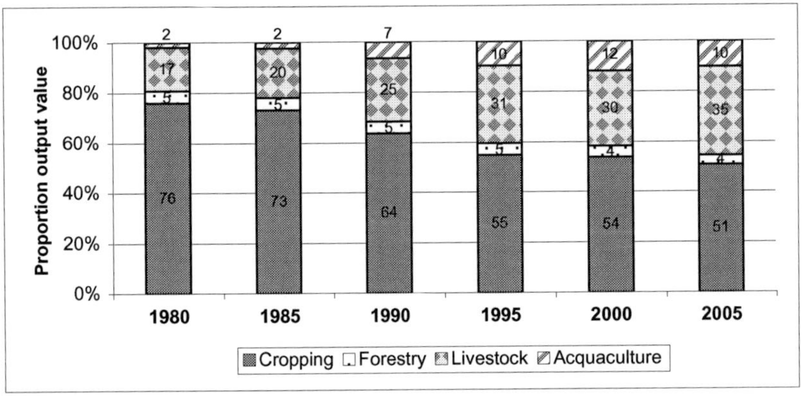

Figure 4.1: Relative output value of major sectors in agricultural economy
Source: All China Data Centre, Accessed August 28, 2008

Figure 4.1 also highlights the relative growth of livestock in the agricultural sector in all periods between 1980 and 2005, except for the period between 1995 and 2000 when output (and therefore output value) statistics were revised down in the 1996/7 Agricultural Census. The output value of the livestock sector grew in absolute terms by an average of 5.5% per year between 1990 and 2005 (All China Data Centre, accessed August 26, 2008). The 11th Five Year Plan (2006 to 2010) targets the livestock sector to increase its share in agricultural output value from 34% in 2006 to 50% by 2010 (Editorial Board of China Agriculture Yearbook, 2006), although the assumptions on which these projections have been made have not been made explicit and appear unrealistic. In parallel with the relative growth of the output value of the livestock sector in the agricultural sector, the share of livestock in total agricultural incomes increased from 16% in 1995 to 22% in 2000 to 25% in 2005 according to Fixed Point Rural Survey data (Zhang, 2006b),[38] although the figure of 30% has been cited by other sources based on National Bureau of Statistics data (China.org.cn, 2006).

The diversification of China's agricultural sector is a response to the restrictions of the central planning era when priority was given a limited number of staple commodities to meet basic human nutrition and warmth (*wenbao*) necessities. Economic development and market reforms allowed for the expression of demand for a greater range of foods and fibres, including a decline in human consumption of grain[39] and an increase in the consumption of meat[40] and other products with positive income elasticities (National Bureau of Statistics, Various years).

The diversification of the agricultural sector is both acknowledged and promoted by a range of policies and programs. One of the early programs in this regard was the Vegetable Basket program that started in 1989,[41]

---

[38] These data derive from the fixed point rural survey system of the Research Centre for the Rural Economy. The survey also reports that the share of income from migration labour in total household income increased from 13% in 1995 to 23% in 2000 and to 32% in 2005.

[39] Between 1983 and 2006, per capita grain consumption for rural residents decreased from 260kg to 206kg, and 145kg to 78kg for urban residents.

[40] Between 1983 and 2006, per capita consumption of meat for rural residents increased from 10kg to 17kg, and 20kg to 24kg for urban residents, although these statistics are drastically understated. Per capita meat consumption is now well above world averages (Waldron, Brown, Longworth *et al.*, 2007, Chapter 1).

[41] The Vegetable Basket program of the Ministry of Agriculture encouraged government at all levels to establish projects and production bases to increase the quantity and quality of non-staple products (including vegetables, fruit, meat and eggs) primarily in city suburbs for consumption by urban residents.

followed by others such as the "Strategic Adjustment of Agricultural Structures" of 1999. In addition to promoting change in the sectoral composition of agriculture, this Structural Adjustment program also aimed to increase linkages between agricultural sectors. For example the Central Committee of the Chinese Communist Party (1998) pronounced that "with the sustainable increase of agricultural production and consumption, it is essential to pay more attention to the livestock industry and promote its links with the farming and processing sectors". It went on to emphasise the social and economic role of livestock in "speeding up the transformation of grain to livestock feed, increasing rural employment opportunities, increasing rural incomes, promoting farming and related industry development and stimulating rural economic development". Livestock has since been singled out for development in all Rural Work Meetings and Number One Documents.[42] All organs and levels of the State have gone about fulfilling the directives through an enormous number of decrees, laws, plans, programs, policies and projects.

4.1.2 Industry composition of the livestock sector

Just as there has been a shift away from traditional mainstay agricultural products like grain in the cropping sector and cabbages in the horticulture sector, there has also been a relative shift out of pork in the livestock sector. While the massive pork industry remains as the pillar of the livestock industry in China, it is in relative decline compared to other livestock industries, as represented throughout meat output statistics in Figure 4.2.

---

42 For example, one of the six major foci of the 1999 Rural Work Meeting was the development of livestock and aquaculture where "income growth will continue to increase the demand for meat which should be met through development of the livestock and aquaculture sectors through strengthening breeding, veterinary and feed grain systems". The 2007 Modern Agriculture program made special mention of livestock development and emphasised safe livestock and meat production, pen feeding in pastoral areas, the development of "small livestock raising areas" (*yangzhi xiaoqu*) in agricultural areas, the expansion of subsidies for small areas, breed improvement for dairy cattle, and the development of disease control and systems.

Along with poultry, sheep meat and dairy, beef has been one of the fastest growing industries in the livestock sector in terms of output.[43]

This change in the industry composition of the livestock sector is also signalled in policy documents. The State Council (1999) released the document titled "Opinion of Accelerating Livestock Development" which stated that over a period of five to 10 years the livestock sector should be restructured by: a) "especially targeting" dairy and fine wool production; b) "promoting" beef cattle, meat sheep and broiler production; and c) "steadily developing" pig and egg production. That is, all livestock industries were targeted for development, but some faster than others. The list was drawn up in the lead up to China's accession to the World Trade Organisation based on detailed projections on the trade balance for these commodities for 2005 and 2015. Findings and "counter measures" were reported by the Ministry of Agriculture (2001). With exception of fine wool production which continued to decline, this targeting has been reflected in actual output figures to 2006. The structural change is targeted to continue into the 11th Five Year Plan (2006–10).[44]

The combination of powerful "demand-pull" and "policy-push" forces causes abrupt output and price cycles in China. The "take off" period can be dramatic. Pork and wool output grew rapidly in the 1980s, followed by poultry and beef in the 1990s, followed by sheep meat and dairy in the 2000s. As the policy attention, "investment waves" and speculative activity subside, this initial growth period is inevitably followed by

---

[43] Other means of estimating the relative growth of industries—through output value and by including the non meat value from livestock—are confounded by changing data collection and reporting methods. In addition to the value of meat from livestock, other products from livestock (*fuchanpin*) include eggs, milk, wool, cashmere, skins/hides and offal. In 2002, these other livestock products represented 19% of the output value of the livestock sector. From 2003, the value of these products was included in the species type from which it was derived. For example, milk and bovine hides were included in the output value from the "bovine industry". This precludes the separation of the output value for bovine meat from other bovine products.

[44] The 11th Five Year Plan forecasts that growth in grain and egg production will stagnate at less than 1% per year. In contrast, the milk industry will be pushed to grow at nearly 8% per year (although this growth ground to halt in the wake of the 2008 "melamine incident"). Growth for the meat sector as a whole is expected to be subdued at 1.6% per year, growing from 78 million tonnes in 2005 to reach 84 million tonnes in 2010. However, the rate will be held back by modest growth in pork and poultry. In contrast, beef, sheep and goat meat combined are expected to increase from 11.46 million tonnes to 13.5 million tonnes at an annual growth rate of 3.6%. Based on the current relative output of bovine and ovine meat output, this is bovine meat production will increase from 7.1 million tonnes in 2005 to about 8.4 million tonnes by 2010.

a period of contraction and correction.[45] The industry then falls back on a more fundamental set of drivers based on underlying market demand, and develops at a more sustainable pace. This process is overviewed for the beef industry in Chapter 5.

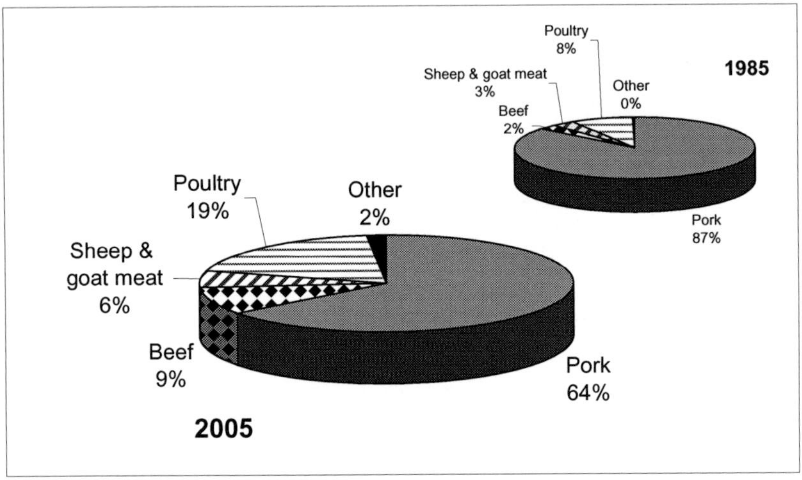

Figure 4.2: Relative output of main meats in China—1985 and 2005
Source: Editorial Board of the China Animal Husbandry Yearbook, 1986 and 2006

4.2 Major forces and challenges facing agricultural industries

While the discussion above looks at change in the composition of sectors and industries, this section examines changes occurring within industries. These are discussed below in terms of industry segmentation, modernisation and integration. All of these forces are part of China's broader modernisation process as indicated by the citations to the "Modern Agriculture" Number One Document of 2007.

45 In the 1990s, virtually all livestock industries suffered periods of over production that resulted in very low prices. Dramatic examples of this occurred in the milk powder industry in 1991–92 and in the egg industry in 1996–97 (Lin, 1997). Pork prices fell dramatically in 1999–2000. The beef industry experienced a sharp market correction in 1996/97, while the lengthy expansionary period in the dairy industry has put severe downward pressure on milk prices to the extent that many producers are unviable. There are signs that the more recently targeted sheep meat industry may be progressing down a similar path.

4.2.1 Segmentation of markets and value chains

The vast majority of Chinese agricultural products are low value and generic in nature and are serviced by equivalent low value agrifood chains. However, Chinese consumers and customers are becoming more demanding, which is reflected in the segmentation of markets for different forms or attributes of a commodity. In the case of beef, market segments are illustrated horizontally across Figure 3.1 and discussed in more detail in Chapter 6. Each of these market segments are linked vertically to agrifood chains that are also undergoing segmentation.

Through this process, China is entering the early stages of industry segmentation that will become more pronounced. As a result of this segmentation, it is becoming less and less relevant for industry, policy makers and researchers to consider agricultural industries and commodities in terms of broad product classes and more and more relevant to consider them as comprised of increasingly segmented product markets and agrifood chains. In simple terms, low value brisket from a culled cow that retails for Rmb15/kg is almost a different commodity to premium grade eye fillet from lot fed steer that retails for Rmb150/kg.

Policy makers are interested in promoting the segmentation process for reasons outlined in Chapters 1 and 2. As an example of policy high level documents related to segmentation, one of the six areas singled in the 1999 Structural Adjustment program is "Upgrading the quality of agricultural products". The program document states that "to meet increasing consumer demands, measures will be taken to increase the variety of products, increase technology and adoption and improve market information, product standard and food safety systems". The 2007 Modern Agriculture program document contained an item titled "Agricultural diversification and integration" which states the need to "develop unique agricultural products in horticulture, livestock and to promote the "one village, one agricultural product" program and village/township specialisation".

A major aspect of the segmentation process is the development of quality and food safety standards, including the Agricultural Products Quality and Safety Law of June 2006. Other recent high level edicts about quality and safety standards include the 2007 Modern Agriculture program document, which contains the item "Develop the market system". It instructs the "continued development of agricultural product quality and safety inspection systems according to food safety laws, the improvement of standards, the improvement of management of agricultural inputs, veterinary products and additives, the strengthening of the production environment, the strengthening of the Non Public Hazard (*wugonghai*) Foods[46], Green Food and Organic Food certification schemes,[47] and the strengthening of registration, trademark and brand management, quality and safety". However as discussed throughout the book and particularly in Chapter 10, while China has issued an enormous number of quality and food safety standards, it has proved much more difficult to bring about adoption and enforcement of the standards to facilitate industry segmentation.

4.2.2 Modernisation of agricultural industries

While segmentation is occurring, the majority of agricultural products in China remain generic in nature and sell into low value markets. Consequently, these markets are serviced by low value agrifood chains as represented on the left hand column of Figure 3.1. The vast majority of participants in Chinese agriculture—in the retail, processing, production and marketing sectors—are still small scale and fragmented, with low technical efficiencies.

China is proactively attempting to modernise these processing, marketing and production structures. This process is seen in China as a process of reducing the relative importance of low value agrifood chains (in the left column of Figure 3.1) and building the relative importance of mid and high value agrifood chains (on the right column of Figure 3.1). This horizontal movement from left to right is thought to increase technical and scale efficiencies, to enable access to higher value markets and to reduce transaction costs. The more centralised structures to improve food safety standards are also argued to provide environmental benefits (e.g. in effluent management)[48] and benefits in the delivery of services (e.g. artificial insemination, vaccination). It is often argued that the larger more centralised structures, especially large agricultural

---

[46] As an example of the size and complexity of these food safety standards, the safety food (*wugonghai shipin*) system has developed 700 national standards for agricultural products, 2,000 agricultural industry standards, and certified 3,500 demonstration standardised areas at province level and 26,000 areas. The scheme relates to pre-production, production and post production stages and essentially aims to guarantee conformity with national standards.

[47] The "triangle" of Green Foods, Organic Foods and Non Public Hazard Foods together account of 90% for all agricultural exports, with Green Foods accounting for 40%.

[48] Under the 2007 Modern Agriculture program, the "Rural infrastructure" item states that "clean resource utilization should be pushed through methane converters from livestock manure and the utilisation of straw for ruminant livestock feed". These are seen to be best achieved by centralising production in "small livestock raising areas". On a related note, under the "Development of organic/biological material" item, mention is made of "increasing the utilisation of straw for fuel, feed and organic fertilizer".

enterprises, have the "power" to access higher value and more distant markets, and can "lead" smaller households into the markets and thereby contribute to rural development. The high value and high technology systems also appeal to policy makers because they represent modernity and nation building.

The emphasis on large scale and high technology structures and enterprises is evident in a large number of official edicts, including the 2007 Modern Agriculture program,[49] the 2002 Agricultural and Rural Work Meeting, the 10th Five Year Plan[50] and the State Council (2003) "Opinion of accelerating the development of agricultural processing".[51] These high level policy documents are an expression of "modernist" thinking and support that is pervasive in Chinese agriculture down to local levels, and supported through an enormous array of policies, programs and infrastructure that is routinely showcased in the media and on official tours. China's agricultural modernisation drive has not been without successes. However, as discussed in this study, "policy-push" support for modernisation has in some cases occurred contrary to "demand-pull" economic settings, and has in some cases distorted and impacted negatively on industry participants.

4.2.3 Integration of industry supply chains

Another aspect of the transition of Chinese agriculture is integration within industry supply chains. This is shown in Figure 3.1 as the vertical

[49] The item "Agricultural diversification and integration" emphasises the need to develop dragon head enterprises through "interest and tax subsidies, subsidies for processing and technology, improving the value added tax system, reducing tax for agricultural processing, and for all levels of government to use public finances to develop investment in agro-industrialisation, develop comprehensive projects, increase loans to dragon heads, especially for working capital for input purchases, and to provide services to dragon heads".

[50] At the beginning of the 10th Five Year Plan in 2001, large scale agricultural processing enterprises generated added value of Rmb697 billion, employed 15.1 million people, exports of Rmb51 billion and profit taxes of Rmb272 billion. The output value was 85% of that of agriculture. The 10th Five Year Plan aimed to increase primary processing of agricultural products to 55% and secondary processing to 35% (Editorial Board of the China Animal Husbandry Yearbook, 2002).

[51] The document states that "over a period of development of five to 10 years, the agricultural processing industry should be established in a way that is conducive to the processing of "superior" agricultural products, should establish key processing enterprises and demonstration bases, technical innovation systems, complete quality and safety standards, and should increase the proportion of agricultural processing industry in gross domestic product and increase industrial adding value".

relationships within agrifood chains, from "downstream" (consumption/retail) sectors to "upstream" (processing and production) sectors and traversing various marketing structures.

Industry supply chains in the central planning era were disjointed by the distribution of jurisdiction across different line bureaus. This was particularly the case in agricultural industries, where eight different line bureaus had jurisdiction (Lieberthal and Oksenberg, 1988; Lin, 1998).[52] The abolishment or demotion of most of these line bureaus in the reform era has simplified the administration of supply chains. However, the administrative reforms have created a void that has not been filled with effective market based mechanisms for integration. This is perpetuated by lack of market institutions including effective market information and effective standards and grading systems, including in the beef industry (Waldron *et al.*, 2003a, Chapter 11).

As a result, the preferences of both customers and consumers are not fully or efficiently passed down the supply chain to producers, and agricultural products are still not produced and assembled in a form required by processors to meet the preferences of these consumers. This lack of integration creates less of a problem in the trade of generic products where there are few customer preferences to relay compared with differentiated products (although markets of course still match overall supply of generic products with demand). As customer and consumer demand become more exacting of the agricultural inputs, closer levels of integration along the supply chain are required, with various implications.

First, there is a widespread belief that increased integration between agriculture and industry provides opportunities in the form of product transformation, value adding, and increased demand and prices for agricultural products.[53] Second, there are counter posing concerns that

---

[52] Most fundamentally, the Ministry of Agriculture and its agencies were concerned with agricultural production and services, the Ministry of Commerce and its agencies were concerned with agricultural marketing and inputs, and the Ministry of Light Industry with processing.

[53] Under the 1999 Structural Adjustment program, the item "Development of agricultural processing industry" states that "To add value to agricultural products and provide off-farm job opportunities, both of which will increase rural incomes, agro-industrialisation through the production, marketing and processing sectors will be encouraged". Processing activities related to livestock (including feed, slaughter, veterinary products, food products, leather, wool and cashmere processing) had an output value of Rmb800 billion in 2005 (Xinhua Net, 2006), nearly 60% of the output value of primary livestock products in that year. The 11th Five Year Plan aims to increase the relative output value from agricultural processing to the output value from primary agricultural products from 1:1 in 2006 to 1.5:1 in 2010 (Editorial Board of China Agriculture Yearbook, 2006).

a lack of integration will devalue agricultural products and lower prices. This is especially the case if intermediate and final customers perceive that they are taking product risks (quality, homogeneity, quantity) that they will seek to offset by discounting the purchase price of the product, which will be passed back to producers. Third, there is a concern that insecure sources of agricultural inputs harm the competitiveness of the industrial sector. Raw inputs make up the bulk of processing costs in many agricultural processing sectors and the price and attributes of the inputs are major factors in the competitiveness of the companies. These agricultural processing companies are often based in rural areas and are an important source of local development, employment, government revenues and incomes for agricultural households selling to the processors. Fourth, there are concerns that a lack of integration will negatively impact on the international competitiveness of Chinese industry and that if Chinese processors and distributors can not obtain the specific inputs they require, they will turn to imports to source their raw materials, as is the case for fine wool (Brown, Waldron and Longworth, 2005) and for higher and more consistent quality beef (Waldron, Brown and Zhang, 2007) and sheep meat (Waldron, Brown, Longworth *et al.*, 2007). Fifth, improved integration is likely to highlight comparative advantages and disadvantages and to increase levels of inter-regional specialisation, which may contribute to the development of more remote areas of China.

China is now taking steps to increase integration in supply chains. On a general level, there have been investments in physical market places,[54] information systems,[55] and as mentioned above efforts to build contract and vertically integrated structures, and quality and food safety standards. While these broad based investments are an important starting point, many of the challenges in increasing integration have to be addressed on a product or industry specific basis, where integration is determined through micro-level

---

[54] Since 1985, the Ministry of Agriculture has supported the development of a network of agricultural product wholesale markets. There are now over 4,200 markets which recorded about US$44 billion in sales in calendar year 2003. The Ministry of Agriculture considers over 300 of the markets to be "major markets" supplying regional areas and major urban centres (USDA, 2005). The Ministry of Commerce (2006) and the Ministry of Finance outline a program to improve the distribution, storage, and quality agricultural products circulation, mainly through substantial subsidies to investors.

[55] For example, the 11th Five Year Plan targeted the development of an information service for nationwide rural markets, and 18 specialised web sites under the Ministry of Agriculture (China News Service, 2006).

activity. Analysis in subsequent chapters, especially Chapter 8, of this study provides some insight into how well or otherwise agricultural industries are integrated.

4.3 Concluding comments

Official aggregate data shows that Chinese agricultural sector is becoming more diversified with the growth of "non staple" industries like beef. The official aggregate data can not however adequately capture other major trends occurring within industries including industry segmentation, modernisation and integration. The policy documents presented in this chapter therefore aimed to show how policy makers seek to promote these forms of industry and agricultural development. These policy settings can have a discernable impact on industry and industry actors, but can also be "policy mirages". The way that these policy settings impact in practice and at a micro level is analysed in detail throughout the remainder of the study.

Chapter Five

Major trends in the beef industry

This chapter draws on official statistics to overview major trends in the beef industry. Official statistics often form the basis for decision making by officials and enterprises and for this reason alone deserve close scrutiny. Statistical inaccuracies and the high levels of aggregation of Chinese official statistics can lead to poor or mis-targeted decision making if not interpreted carefully. In recognition of the shortcomings in Chinese official statistics, China has endeavoured in recent years to increase the accuracy of the official statistics, collect new data series and to build reporting systems.

Another reason why official statistics are analysed in this chapter is to provide context for subsequent chapters and to supplement the more detailed analysis presented in Chapters 6 to 9. In this regard, official statistics are useful for several reasons. First, the statistics are the only source of information for comprehensive nationwide data and therefore provide context for the more specific or representative case study approaches. Second, some of the data (for example scale of production data) can be cross referenced against budgeting results to explain trends in the industry. Third, the official production data in particular can be analysed to derive a series of technical indicators (such as carcass weights) which are incorporated into some aspects of the budgeting analyses in Chapters 7 and 9.

Even the incomplete official statistics provide some indication of the way that the Chinese beef industry is developing, commercialising and maturing. The household and marketing reforms of the late 1980s and government policy in the 1990s led to the rapid growth of the beef industry to lay what is known in China as the "production base" for the industry, with an emphasis on expansion of the cattle herd, and an emphasis on quantity or output over quality or efficiency. Due to the changing incentives of households—especially the increasing opportunity costs of labour and the mechanisation of cultivation—the beef cattle herd is no longer expanding and in fact began contracting in more developed parts of China in the 2000s. However, the industry is becoming increasingly oriented to the production of the primary output from the cattle, namely beef, in a process known in China as industry commercialisation, which is a key component of the overall modernisation process.

5.1 Reconciling official production and consumption statistics

The major inaccuracies in official Chinese statistics and in particular livestock statistics are widely reported (Fuller *et al.*, 2000; Ke, 2001; USDA, 1998; Ma *et al.*, 2005; Longworth *et al.*, 2001, Appendix A). Most fundamentally the large discrepancy between production (which is overstated) and consumption (which is understated) can not be explained through net trade. These and other issues associated with inaccuracies, conversion and aggregation in the official statistics for the beef industry are depicted in Figure 5.1.

On the production side, the extent of overstatement of production data was highlighted in the National Agricultural Census of 1996 and 1997 when livestock and meat production were drastically revised downwards following the census. The largest revisions were made for bovines and bovine meat. The extent of over statement for bovines and bovine meat were a reflection of the emphasis that policy makers had been placing on the development of the industry in the 1990s. Bovine numbers were revised down by 16.5% and bovine meat by 28.1% for 1996 (USDA, 1998). Since the revision, output statistics for livestock have again been overstated by, it was believed by industry officials, around 20% (Animal Husbandry Bureau, personal communication). In fact, in the wake of the Second National Agricultural Census statistics were revised down even more dramatically. For example, in 2006 bovine numbers were revised down by 25% and beef output by 23% (China Animal Husbandry Yearbook, 2009).

On the consumption side, official statistics are drastically understated, mainly because they only account for in-home consumption. In cities like Beijing, it has been claimed that out-of-home beef consumption may be as high as 40% of total beef consumption (Beijing Animal Husbandry Bureau, personal communication). The failure to take into account increasing amounts of out-of-home consumption especially in urban areas means that not only are absolute consumption levels understated, but that the levels become more understated over time. In addition, beef consumption is estimated on a bone-out basis while beef production which is estimated on a bone-in basis. It is estimated in China that one unit of bone-in carcass weight is the equivalent of 0.73 bone-out beef weight, although there are large differences between regions, breeds and cattle production systems.

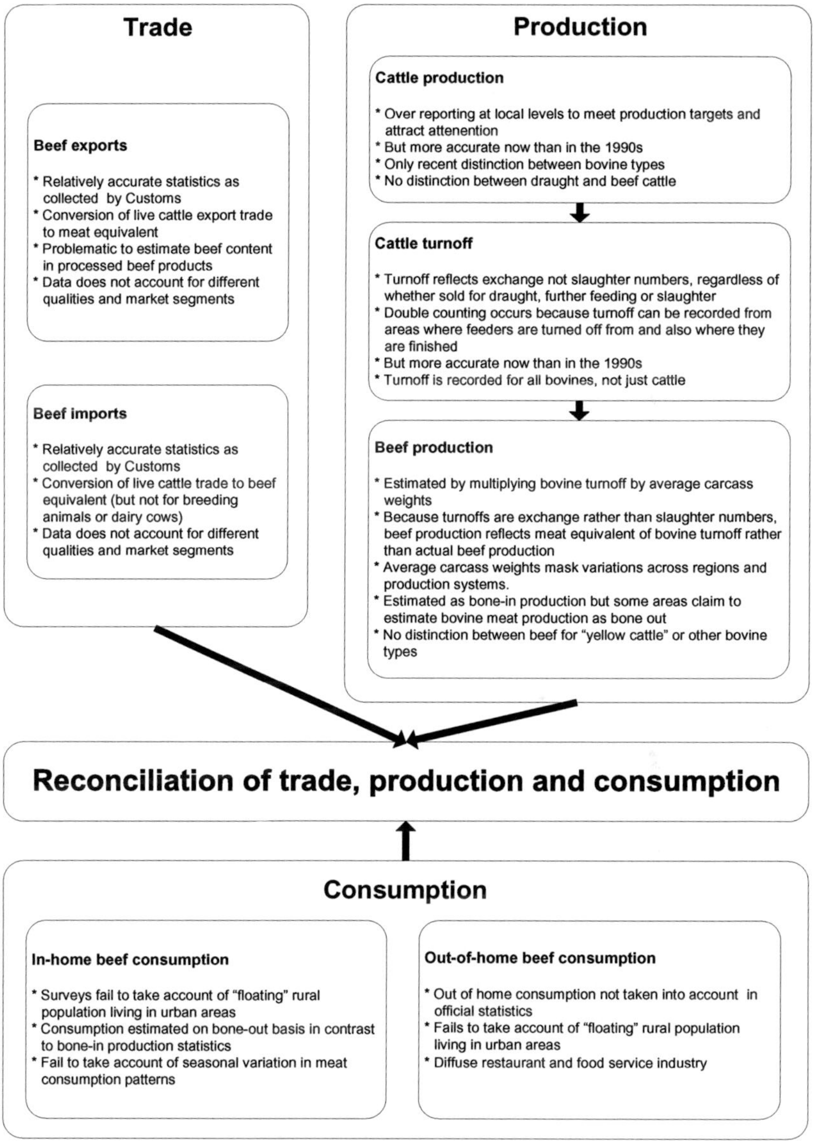

Figure 5.1: Issues in official statistics for the Chinese beef industry

5.2 Consumption indicators

Official aggregate consumption statistics are derived from both the Urban Survey Team and the Rural Survey Team of the National Bureau of Statistics. The official beef consumption figures based on consumer surveys for urban and rural households are shown in Figure 5.2, which also includes an alternative method of determining beef consumption, namely the "trade balance method", where consumption is derived from domestic production minus net trade divided by population.[56] This is the equivalent of "per capita supply" used in FAOSTAT. Figure 5.2 also estimates consumption (derived from the "net trade method") on a bone-out basis.

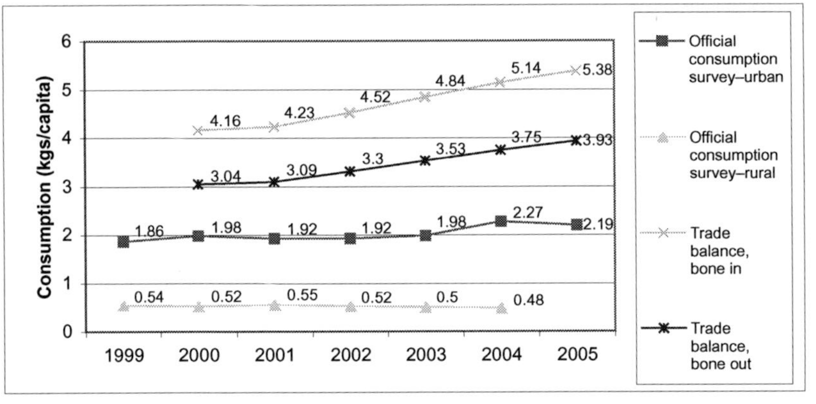

Figure 5.2: Beef consumption estimates based on two different sources of official data
Source: China Statistical Yearbook, China Animal Husbandry Yearbook, China Customs Yearbook

The trade balance method provides a more accurate indication of beef consumption levels than the official data. As a reflection of this, estimates suggested by the trade balance method are consistent with those of surveys such as Wang *et al.* (2004). However, a major qualification is that the trade balance method is based on overstate production figures. Another drawback of the trade balance method is that it only allows for estimates of consumption on a national level. The official consumption data provides

---

[56] The net trade component accounts for imports and exports of beef, processed beef products and beef on live cattle, for which data is available in the Animal Husbandry Yearbooks. For details on application of the method to beef and sheep meat see Longworth *et al.* (2001, Appendix A), Waldron, Brown, Longworth *et al.* (2007, Section 2) and Waldron, Brown and Zhang (2007, Chapter 2).

more disaggregated indicators on regional variations, differences between urban and rural consumption, and income effects for in-home consumption.

As mentioned earlier, the large gap between the official consumption figures and the trade balance method figures is due mainly to the large and growing amount of beef consumed out-of-home in the hotel, restaurant and institution (HRI) trade. This is particularly the case in urban areas where most out-of-home consumption takes place.[57] As discussed in more detail in Chapter 6, beef plays a role in an increasingly diversified Chinese menus (such as "sizzling iron beef"), in popular minority and regional dishes (hot pot, Xinjiang skewers, Hui Muslim noodles, Korean dishes) and in foreign restaurants in China (Japanese/Korean barbeque, Brasilian beef, American fattened beef), where beef is the major item on menus

Regardless of the method used to estimate consumption, the average per capita consumption of beef in China of less than 4kg per year is low by international standards. The trade balance method suggests a high growth in beef consumption (at around 6% per year), although this by definition is a function of the increase in beef production (given little change in net trade). If this is adjusted to take account of overstatements in data on the production side (around 20%), then the annual growth rates in consumption might be in the order of 4.8% per year. In this case, the official data suggests that only half of this increase (2.5%) can be accounted for by in-home consumption by urban consumers between 1999 and 2005, and the other half from out-of-home consumption.

The official consumption data indicates that the average annual increase of in-home beef consumption between 1999 and 2005 (of 2.5%) was similar to that for all meat (3%) and therefore the proportion of beef in the total in-home meat diet of urban consumers varied little (at around 7%). Beef consumption grew faster than for pork (2%) and sheep and goat meat (1%),[58] but was lower than for poultry (7%). Change in annual consumption levels for beef have been less volatile than for other meats such as pork, poultry and dairy, many of which have undergone short term fluctuations due to disease outbreaks or food safety concerns.

---

57 The Beijing Animal Husbandry Bureau estimates that of the official beef consumption in Beijing (3.8 kilograms per person per year), an additional 40% is consumed out of home. Ma *et al.* (2006) and Gale and Huang (2007) also find that more meat is consumed out of home than any other food group. Gale and Huang (2007) find that while the lowest income percentile group spends only 8% of its food expenditure out of home, the top income percentile groups spends 30%.

58 Because a lot of sheep and goat meat is eaten out of home in hot pot restaurants, there is a larger discrepancy for sheep and goat meat than for beef between the official consumption statistics and the trade balance method.

Official consumption data indicates that in-home urban beef consumption is at least four times that of in-home rural consumption. The difference between urban and rural beef consumption is likely to be even larger if out-of-home consumption is taken into account. Rural beef consumption averages only around 500 grams per person per year and has actually declined slightly since 2001. China's rapid urbanisation may therefore have an impact on beef consumption as examined in Rae (1998) and Ma *et al.* (2004).

On a regional basis, the Northeastern provinces (Jilin, Liaoning, Heilongjiang), the Western pastoral areas (Inner Mongolia, Qinghai, Ningxia, Xinjiang, Tibet) and the Northern urban areas (Beijing and Tianjin) have relatively high levels of beef consumption. This conforms to widely held perceptions that beef is traditionally eaten by northerners in cold areas. Highly populated provinces in China like Henan, Sichuan and Shandong—the three largest beef producing provinces in China—have very low levels of beef consumption amongst both rural and urban populations.

The relationship between incomes and in-home urban beef consumption in the official data is shown for 2002 in Figure 5.3. Growth in both per capita beef consumption quantities increase as incomes increase until incomes becomes "high",[59] at which point marginal growth rates decline. This conforms to survey results for beef from Longworth *et al.* (2001, Chapters 13 and 14), while Gale and Huang (2007) find that income elasticities for beef and sheep and goat meat diminish as income increases. Thus, the consumption of beef becomes saturated with increasing incomes, as beef makes up a part—but not an especially important part—of an increasingly diversified diet.[60]

The statistics above suggest that the growth in levels of in-home beef consumption has been consistent but modest. Looking into the future, significant growth in out-of-home beef consumption may continue, but this is dependant on the continued popularity of hot pot and other culinary trends. As an indicator of possible beef consumption trends in some parts of mainland China, the high income and highly urbanised Chinese island of Taiwan had a per capita beef consumption of around four kilograms between 2000 and 2006 (USDA, 2006).

---

[59] The seven categories of income levels relate to incomes percentiles, where "high" relates to the wealthiest 10% of the urban population.

[60] This pattern also applies (to varying degrees) with other most other livestock products with the exception of aquatic products (which has the highest income elasticity) and pork (which has the lowest) (Liu *et al.*, 2007; Meng, 2007; Wang *et al.*, 2004). Meat protein replaces food grain and vegetables as incomes increase (Wang *et al.*, 2004; Rae, 1997; Ma *et al.*, 2004), while oil intake remains constant.

This circumspect perspective of beef consumption in China is contrary to that held by many agribusiness concerns and policy makers—both Chinese and foreign—that with urbanisation and income growth, Chinese consumers will adopt beef consumption levels and patterns more like Japanese, Korean or Western consumers. Policies based on this broad assumption are not well supported and, as discussed below, have often led to inappropriate policy and investment decisions.

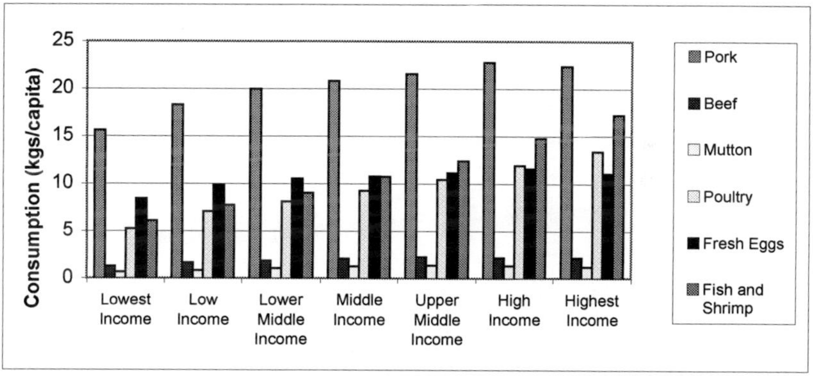

Figure 5.3: In-home consumption of animal products and income strata in urban areas—2002
Source: China Statistical Yearbook, 2003

5.3 Price indicators

The largest and longest series of official price data for livestock products is collected by the Ministry of Agriculture and reported in numerous places including the *Peasants Daily*, agri.gov.cn, and the *China Animal Husbandry Yearbook*. Prices collected over a large number of sales daily from "observation points" throughout China are averaged out over monthly periods to province level and to represent the monthly average prices for generic types of livestock products, including bovine meat. There are some major limitations to these data, especially the lack of differentiation between types of beef or indeed types of bovine meat. However, since the vast majority (probably 80–85%) of the beef in China is not differentiated by form of product so the data provides some insights into developments in the beef market.

Figure 5.4 provides a simple graph of price levels in one province of each of the five regions to demonstrate several issues. First, generic beef prices dropped substantially in the latter half of the 1990s. The price drop occurred earliest (in 1996) in the Northeast and the Central Plains, but was followed with a less dramatic reduction and at a later stage in the Southwest and Northwest. This reflects a correction from the rapid increase in production in the former two zones in particular in the 1990s without the corresponding increases in consumption. Since 2000, the prices for all provinces have increased steadily, reflecting the more balanced production levels and a steady increase in consumption. Prices for beef increased significantly after this period, especially in 2007 and 2008, to reach a national average of Rmb27.5/kg for August 2008 (agri.gov.cn, accessed August 27, 2008).

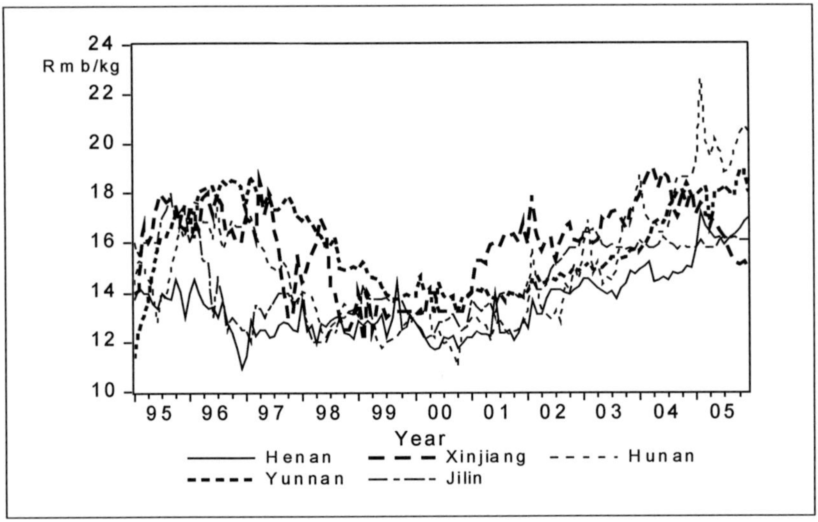

Figure 5.4: Monthly beef prices in selected provinces—1995 to 2005
Source: China Animal Husbandry Yearbook, various years
Note: Prices are expressed in Rmb per kg of generic beef and are not deflated against a base year. Each of the five provinces falls into one of the five beef zones discussed in Section 5.5, i.e. Henan (Central Plains), Xinjiang (Northwest), Yunnan (Southwest) and Jilin (Northeast). Hunan has traditionally been classed within the "Other" zone, but development of the provincial industry in recent years enables it to be classed within a defined beef zone, namely the "Southwest" beef zone.

With regard to regional variations, Figure 5.5 shows average provincial beef prices for one year (2005). Absolute price levels vary considerably across China. Beef prices generally reflect the availability of cattle and beef locally. The lowest prices were recorded in and around Beijing (especially Hebei and

Shandong). The highest prices were recorded for the east coast provinces south of Shanghai. High prices were also recorded for Southwest with the exception of Sichuan which has a large draught cattle and buffalo population.

More detailed co-integration analysis reported in Brown *et al.* (2002b) and Waldron, Brown and Zhang (2007, Section 2.6) shows close integration in generic beef prices between provinces within beef zones (see Section 5.5 below), suggesting that cattle production within the zones are similar and that beef is readily transported across provinces. However, the extent of integration across the zones is much less pronounced, reflecting different production systems and that beef is not easily transported over long distances. This is certainly the case for the fresh beef that dominates the beef market in China. In addition the price differences between provinces appear to be higher than suggested by transport differentials alone.

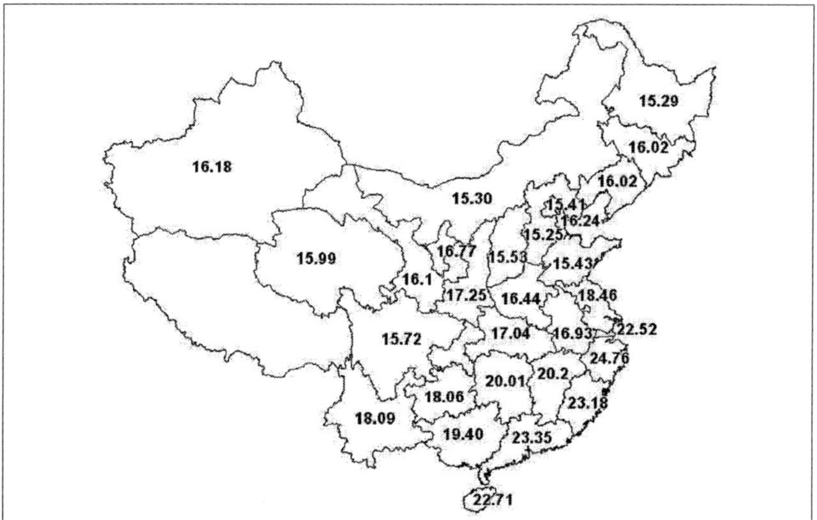

Figure 5.5: Average generic beef prices by province (Rmb/kg)—2005
Source: China Animal Husbandry Yearbook, various years

Figure 5.6 shows the relationship between prices of selected meats. The data shows that bovine meat prices are closely correlated with sheep and goat meat prices, but less correlated to pig meat prices. Beef and sheep/goat meat prices show less inter-year variation than pig meat prices (and to lesser extent poultry and eggs) as monogastric livestock are more affected by disease outbreaks and feed prices. This is further illustrated (though only visually) in Figure 5.7, which suggests that pig meat prices follow corn prices more closely than do bovine and sheep and goat meat prices. Feed prices increased significantly in 2007 and 2008 to reach, for example, Rmb1.76 in Shandong in August 2008 (agri.gov.cn, accessed August 27, 2008). While concentrate feed such as corn is not a major input for most of the cattle produced in unspecialised household systems, the relationship between feed and cattle and beef prices is highly relevant in more intensive production systems analysed in Chapter 9.

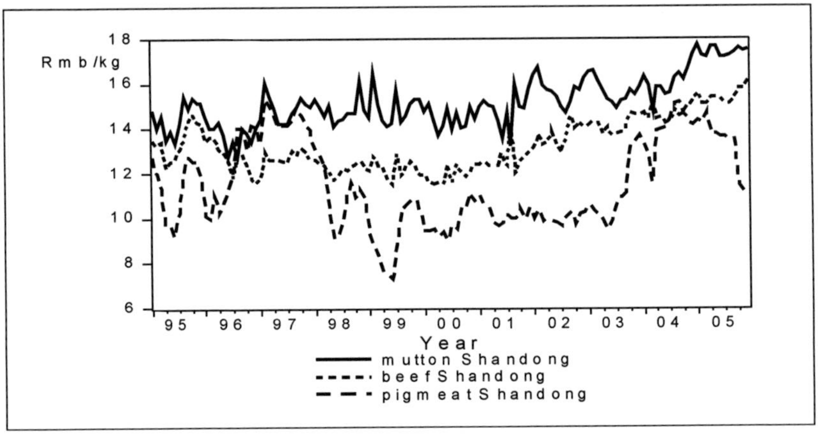

Figure 5.6: Monthly meat prices in Shandong—1995 to 2005
Source: China Animal Husbandry Yearbook, various years

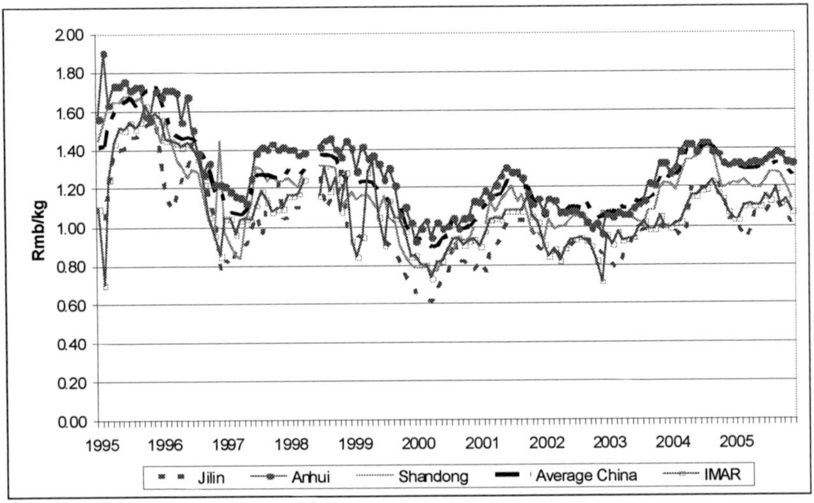

Figure 5.7: Monthly corn prices in selected provinces—1995 to 2005
Source: China Animal Husbandry Yearbook, various years

As mentioned, price data above does not cover the large price increases of up to 20% that occurred in 2007 and 2008 for all grain, meat and other agricultural products. These developments could be expected to impact on different sectors and participants in the beef industry. However, as discussed for pork (Nan and Han, 2008), broilers (Zhai and Han, 2008) and dairy (Li *et al.*, 2008), the price increases for livestock products have often been offset by the costs of inputs and increases in the margins of non-farm actors in supply chains. While the analysis in Chapters 6 to 9 does not specifically analyse change in price parameters in 2008, it does include examine sensitivities to various price relativities. Price relativities rather than absolute price levels are the most important determinant of economic incentives for industry actors.

5.4 Production indicators

Production indicators show that after a policy led boom in production in the early stages of industry development, a correction occurred which was followed by more subdued industry growth. From a low base, technical efficiencies, levels of commercialisation, household specialisation and regional specialisation are increasing.

5.4.1 Bovine, bovine turnoff and bovine meat production indicators

Major production indicators for the beef industry are presented in Figure 5.8. Between 1979 and 2005 all bovine related production indicators—bovine numbers, turnoff and bovine meat—increased at high rates compared to most other livestock industries and in international terms (see Waldron, Brown, Longworth *et al.*, 2007).

Growth rates were especially high in the 1990s when the industry was aggressively pushed by government and economic actors attempting fill a perceived shortfall in supply and to build a "production base" for beef. This was orchestrated under the Straw for Ruminants program, as discussed in detail in Waldron *et al.* (2003a, Chapter 9). In the fervour to meet targets and generate political favour, local government officials dramatically overstated output figures in the period by up to 28%, which led to the statistical correction following the Agricultural Census in 1996 and 1997. Over-reporting and subsequent correction occurred in all livestock industries, but was highest in the beef industry.

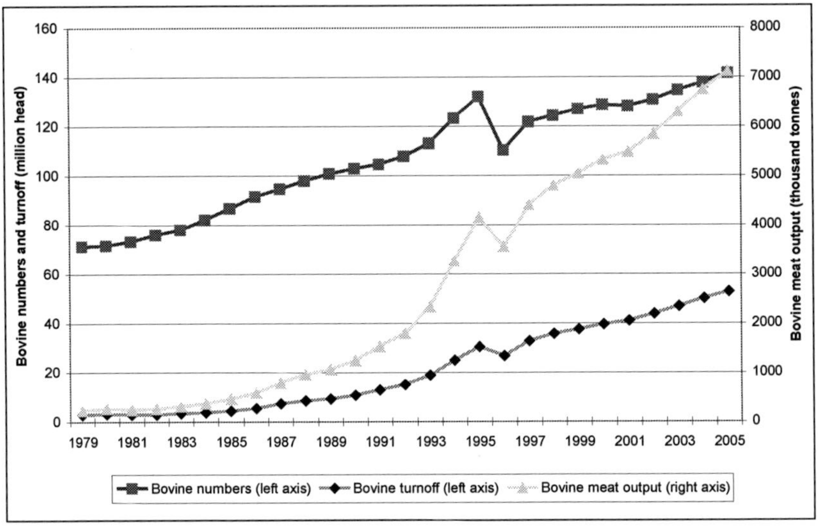

Figure 5.8: Bovine and bovine meat production indicators—1979 to 2005
Source: China Livestock Yearbook, various years
Note: Bovines refer to beef cattle (referred in Chinese statistics as "yellow cattle"), dairy cattle, buffalo and yaks

While there has been a continuation of the Straw for Ruminants program and other programs such as the "Favoured Area" program (Waldron, Brown and Zhang, 2007), policy directives aimed at expanding output have subsided and market constraints and underlying economic forces have set in. The correction to the over expansion of the 1990s is indicated in Table 5.1, which shows a slowdown in growth for all output items in the 2000s. As mentioned in Chapter 4, this is a common pattern in all livestock industries.

The slowdown in growth is even more significant if two further factors are taken into consideration. Growth rates for some of the indicators to 2005 would be negligible or negative if official statistics are revised down by about 20% to take account for estimated statistical inaccuracies, as mentioned above. Furthermore, the actual official production indicators for 2006 and 2007 suggest that the rate of increase in these years was subdued.

Table 5.1: Average annual growth rates in cattle and beef production by five-year periods

5 year period	5 year plan	Era	Bovines	Bovine turnoff	Bovine meat
1980–85	6th	Household responsibility system	4%	10%	18%
1986–90	7th	Market reforms	2%	19%	23%
1991–95	8th	Industry commercialisation, Straw for Beef	5%	27%	34%
1996–2000*	9th	Straw for Beef, agricultural census	3%	10%	10%
2001–05	10th	Rationalisation and stabilisation	2%	6%	6%
2006–10**	11th	Modernisation	—	—	3.6%**

Source: China Livestock Yearbook, various years
Data for the 1996–2000 period is based on revised figures after the Agricultural Census for 1996/7, which is also a factor in the lower output statistics.
** The 2006–10 figure is projected under the 11th Five Year Plan for livestock.

5.4.2 Cattle production

The slowdown in industry growth is most pronounced in the bovine and cattle production sector of the broader beef industry. Between 2000 and 2005, annual growth in bovine numbers averaged 2% to reach 142 million head. However, much of this growth in bovine numbers was from dairy cattle (to reach 12 million in 2005) and yaks (to reach 7 million[61]), while buffalo numbers remained constant. The growth in numbers of beef cattle (called "yellow cattle and their crosses" in China) was negligible between 2000 and 2005, and actually began reducing from 2004 to reach 100 million head in 2005 (as indicated in Figure 5.9) and 98.6 million head in 2006. The contraction in the yellow cattle herd size was most pronounced in developed beef provinces. Between 2000 and 2006, the number of yellow cattle fell by 25% in Shandong and 48% in Anhui. A similar pattern applies to the number of productive breeding female beef cattle which have also stagnated for China as a whole at around 44% of the national beef cattle herd. Between 2000 and 2006, the number of productive breeding females decreased by 16% in Shandong and 40% in Anhui. The data on cattle numbers is consistent with fieldwork observations, interviews and budget analysis that show that households are culling unproductive and uneconomic breeding females and that draught cattle are being replaced by mechanised cultivation and transport equipment.

[61] The published figures on yaks are said by industry experts (Zhang Cungen, personal communication) to be residual statistical items and are not reliable.

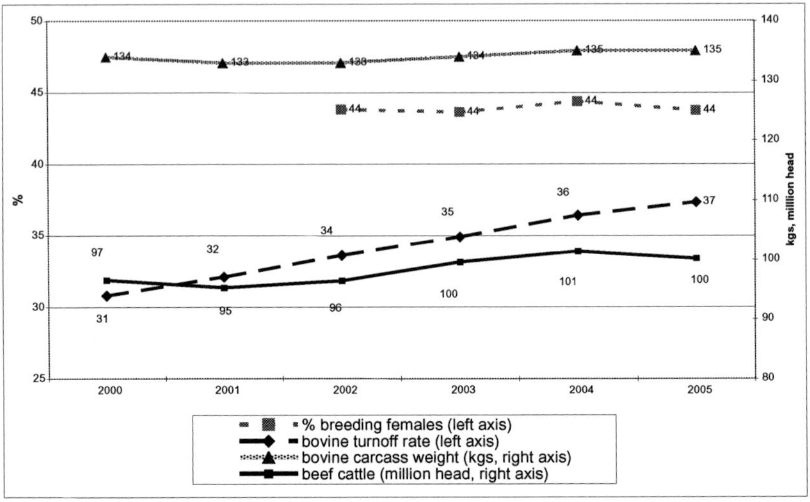

Figure 5.9: Beef cattle numbers and productivity indicators—2000 to 2005
Source: Derived from China Livestock Yearbook, various years

Another major trend in the cattle production sector is a switch out of small scale unspecialised cattle production toward larger scale specialised cattle feeding. This trend is partly discernable in scale of production data presented in Table 5.2 which shows a number of points.[62] First, cattle production in China is dominated by small scale unspecialised cow–calf (breeding) households. However, nearly two million of these households exited this very small scale production sector (1 to 9 head turnoff per year) between 2003 and 2005. The proportion of cattle turned off in this scale category decreased by 7% and the average number of cattle turned off by households in the category increased from 2.3 to 2.5 head. Second, this reduction in the small scale production sector was complemented by a gradual increase in the number of households and cattle turned off from the mid sized units (10–49 head, 50–99 head, 100–499 head), which indicates that household specialisation in

[62] Scale of production information has become publicly available in the China Livestock Yearbook since 2000. For scale data before that period, see Longworth *et al.* (2001, Chapter 4) where data was collected by the authors. While scale data has been collected since 2000, it is not possible to analyse change this far back because scale categories and units (number in stock versus number turned off) have changed from year to year. The data includes provincial breakdowns but regional trends are difficult to discern because of the importance of particular prefectures within a province. For a full discussion of the scale data and regional trends see Longworth *et al.* (2001) and Waldron, Brown and Zhang (2007).

Table 5.2: Scale of beef cattle production in major provinces in China, 2003 and 2005

	Turnoff 1-9 head		10-49 head		50-99 head		100-499 head		500-999 head		1000 over	
	No. units	% turn off	No. units	% turn off	No. units	% turn off	No. units	% turn off	No. units	% turn off	No. units	% turn off
Guangxi												
2003	638,609	95	2,635	4	250	1	17	0		0		0
2005	606,744	91	4,444	6	486	2	117	1	5	0		
Guizhou												
2003	1,026,470	94	5,742	5	160	1	15	0		0		0
2005	992,120	90	8,226	8	172	1	73	1		0		0
Anhui												
2003	873,135	90	6,145	8	497	1	82	1	10	0		0
2005	743,625	88	5,405	8	950	3	59	1	1	0		
Sichuan												
2003	1,422,252	90	6,999	6	989	2	205	2	15	0	1	0
2005	805,505	86	7,672	9	692	3	102	1	6	0	3	
Henan												
2003	2,886.817	84	31,424	9	3,379	4	503	2	39	0	10	0
2005	2,777,432	86	27,497	9	2,610	3	515	1	26	0	6	0
Shaanxi												
2003	366,571	83	4,568	12	283	2	37	1	8	1	4	1
2005	463,068	84	4,577	11	229	2	49	2	6	0	3	
Yunnan												
2003	596,392	91	6,188	6	433	2	95	1	6	0		0
2005	706,142	83	11,051	12	774	3	233	2	2	0		
Hunan												
2003	902,475	81	15,472	13	1,2 44	5	48	1	3	0	2	0
2005	750,503	79	13,894	16	1,322	4	121	1	3	0	1	0
Ningxia												
2003	173,833	69	3,119	22	231	5	27	2	3	1	3	2
2005	198,355	77	4,193	15	184	3	58	3	6	1	2	2
Jiangsu												
2003	145,443	89	1,417	6	164	2	21	1	4	1	3	1
2005	119,273	76	1,485	12	329	7	47	4	6	1	1	
Shandong												
2003	1,268,167	63	29,042	29	3,603	6	469	2	42	1	7	0
2005	1,251,111	72	29,243	16	4,801	7	600	4	81	1	16	
Shanxi												
2003	185,021	76	4,674	14	563	6	113	3	10	1	3	1
2005	179,699	69	5,894	19	724	7	176	5	7	1	2	2
China												
2003	**15,536,964**	**72**	**386,211**	**16**	**42,742**	**5**	**8,760**	**4**	**1,072**	**1**	**289**	**1**
2005	**13,617,317**	**65**	**410,688**	**19**	**56,379**	**8**	**11,238**	**5**	**1,073**	**1**	**200**	**1**
Hebei												
2003	1,061,583	60	49,377	25	5,342	7	1,404	6	86	0	16	2
2005	995,520	56	54,256	28	5,635	7	1,264	6	65	1	17	3
IMAR												
2003	178,554	54	29,538	26	3,091	9	566	4	72	3	38	3
2005	345,650	54	48,565	24	4,952	11	1,298	6	140	3	25	1
Gansu												
2003	329,419	68	11,328	15	2,068	13	130	2	18	1	1	0
2005	458,227	48	14,364	41	1,926	9	110	1	7	0		0
Xinjiang												
2003	245,740	63	6,503	14	2,793	12	607	6	56	2	19	2
2005	275,250	46	22,243	27	5,180	20	661	4	78	3	11	2
Heilongjiang												
2003	430,836	46	63,497	35	4,785	13	980	5	72	2	4	0
2005	200,928	45	25,318	33	4,839	13	699	7	68	2	5	5
Liaoning												
2003	490,711	41	41,357	24	5,147	14	1,173	11	207	6	44	3
2005	557,160	41	38,387	27	7,398	18	1,172	9	136	3	22	3

Source: China Livestock Yearbook, 2004 and 2006

Note: Some provinces are omitted form the table due to small cattle numbers (Hainan, Fujian, Guangdong, Zhejiang) and apparent statistical anomalies (Chongqing, Hubei, Jiangxi, Tibet, Qinghai, Jilin) the industry. Third, however, there are few genuinely large feedlots in China (that turn off over 500 head) and their numbers are decreasing. Fourth, the scale of production tends to be lowest in the more remote and less developed areas of China and increases in the more eastern and well developed parts of China. The qualification is that scales of production are relatively high in the northwestern provinces (due to pastoral or semi-pastoral systems) and in the northeast (due to grain supply and prices for larger feeding operations). Other major regional variations are discussed in Section 5.5 below.

5.4.3 Cattle turnoff

While the growth of bovine numbers declined and the absolute number of beef cattle declined, both bovine turnoff numbers and bovine meat output increased by an average of 6% per year between 2000 and 2005. Although lower than previous eras, the increase in bovine turnoff and bovine meat production outpaced the growth in bovine numbers by a factor of three.[63] While this partly reflects the culling of cows, it is also a reflection of the increasingly efficient conversion of cattle into beef, or what is referred to in China as industry "commercialisation" or "commodification" (*chanpinhua*).

The turnoff rates for China as a whole and in selected areas are shown in Figure 5.10. The increase in turnoff rates in particular has been generated by the culling of unproductive females, a shift from the production of cattle for draught purposes to beef production purposes,[64] higher calving rates and the uptake of breed improvement services.[65] This is reflected in increasing levels of scale and specialisation in the cattle production sector (as discussed above) and economic incentives to change production systems both within the units (as discussed in detail in Chapter 9).

---

[63] This trend is likely to be more pronounced than reported given that much of the increase in bovine numbers was due to increases in dairy cattle numbers, which have a lower turnoff rate than yellow cattle.

[64] Agricultural mechanisation of the 1980s and 1990s has facilitated a shift in the function of cattle from draught to beef. For example, total agricultural machinery power increased from 14,746 kW in 1980 to 38,547 kW in 1996, while the number of mini tractors increased from 1.9 million to 9.2 million over the same period (Hu *et al.*, 1998).

[65] The reach of the breeding system is evidenced in the breed improvement rates. Only 30% of cattle in Shandong were improved in 1998, compared to 40% at the end of 2004 (Shandong Animal Husbandry Bureau, 2005). However, interviews with local officials record a large range of AI coverage, with very high rates in particular areas (e.g. 90% in Dezhou in 2006) and very low rates in others (20% in Southwestern provinces such as Sichuan and Hunan) because the larger framed improved cattle are not well suited to mountainous areas.

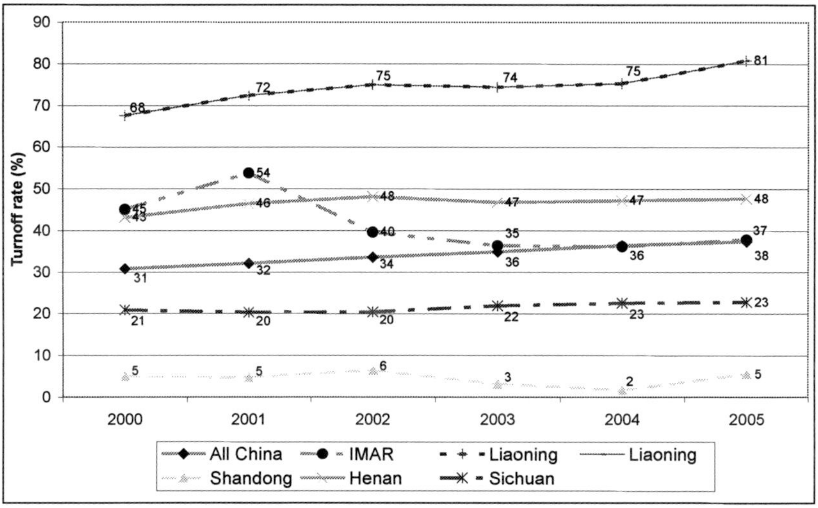

Figure 5.10: Turnoff rates for bovines in China and selected provinces—2000 to 2005
Source: Derived from China Livestock Yearbook, various years

Another aspect of turnoff is the destination and form of cattle turnoff. Turnoff statistics include both cattle that are slaughtered after sale and also cattle that are traded for further feeding or transport to other areas.[66] In regions like Inner Mongolia, pastoral households also slaughter cattle for self consumption. A substantial and increasing number of cattle in China are not slaughtered immediately but instead on-sold for further feeding locally or in other areas. While systematic data on turnoff destination for China as a whole are not available, these are available in the case of Inner Mongolia as shown in Figure 5.11, where 27% of cattle turned were destined for trading or further feeding.[67] This figure is larger than the national average (estimated by industry officials as about 10%) but lower than areas like Jilin, Liaoning and Shandong where there is a large feeding sector. These cattle

[66] This causes major statistical errors, as the turnoff numbers have historically been treated as slaughter numbers, which is then multiplied by an average carcass weight to estimate beef production. Officials have sought in recent years to take this into account when estimating slaughter numbers. However, it is difficult for even county level officials to estimate cattle slaughtered or cattle traded. Slaughter numbers and beef production is not able to be recorded at slaughter level in China because of the highly fragmented slaughter structures.

[67] These figures for Inner Mongolia are skewed by the fact that it has a large dairy herd (nearly 50%) and that is a major source of feeders for other parts of China through large live cattle markets on the eastern grasslands.

are destined for intensive feeding and slaughter counties such as Yangxin County in Shandong which buys in more than double the number of cattle that are in stock at the end of the year and therefore has turnoff rates of more than 200%. The increase in turnoff for further feeding and inter-regional trade is likely to become more pronounced in the future.

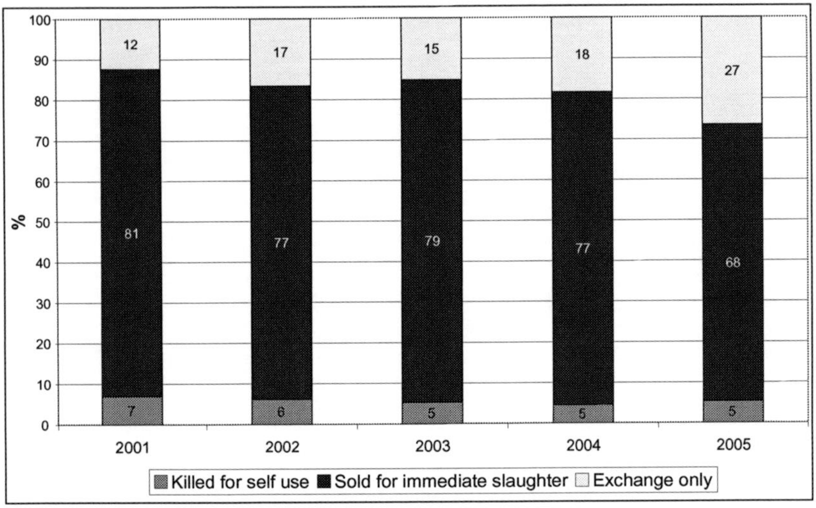

Figure 5.11: Cattle sales and slaughter channels in Inner Mongolia
Source: Inner Mongolia Statistical Yearbook, various years

5.4.4 Beef production

While bovine meat production increased faster than turnoff from 1980 to 1996, the two items have grown at roughly the same rate since 1996. This is due to stagnation in the carcass weights from Chinese cattle at just 135 kilograms bone-in, as shown in Figure 5.12. This does not necessarily indicate stagnating technical levels, such as breed improvement and feed conversion ratios. Rather, Chinese cattle producers—households in particular—are selling cattle off at a younger age because they do not have the economic incentives to feed for longer periods, as shown in Chapter 9. The implications are that Chinese consumers—even in mass markets—are able to access beef from relatively young cattle, which improves quality attributes such as tenderness and colour. However, these young, light cattle are not suited to premium beef markets (as a finished product) where larger cuts are required from larger animals that must be produced through feedlot systems.

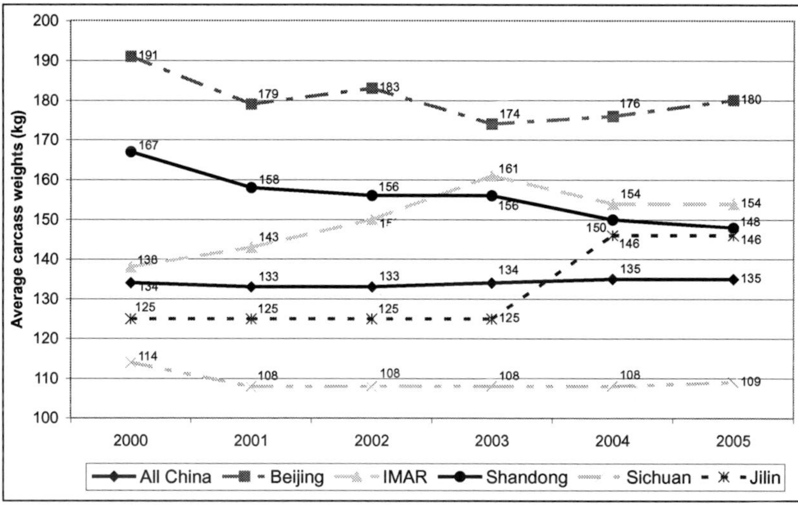

Figure 5.12: Average bovine carcass weights in China and selected provinces—2000 to 2005
Source: Derived from China Livestock Yearbook, various years

According to projections under the 11th five year plan, the beef industry is expected to slow further over the period 2006–10. Bovine, sheep and goat meat combined are expected to increase from 11.46 million tonnes to 13.5 million tonnes at an annual growth rate of 3.6%. If so, and if the proportion of bovine meat output to sheep and goat meat output remains as it is at present, then bovine meat output will be 8.4 million tonnes by 2010.[68]

5.5 Regional production indicators

Within the broad national trends in the industry above, there are a series of important trends occurring at the regional and inter-provincial level. At the most basic level, the distribution of bovines by province is shown in Figure 5.13. The geographic distribution of the beef industry can be divided in regional zones, provinces, and "pockets" (prefectures or groups of prefectures). For detailed analysis of the regional distribution of the cattle

[68] Simpson (2003) projects that under a "robust" growth scenario that beef cattle numbers will increase to 123 million head in 2010. The USDA (2007) projects bovine slaughter at 66 million head, beef production at 9 million tonnes, imports at 8,000 tonnes, exports at 89,000 tonnes and (as a result), domestic consumption at about 9 million tonnes.

and beef industry, trade, integration and markets see Brown *et al.* (2002b), while Smith (2001) analyses inter-regional flows of cattle and beef in China.

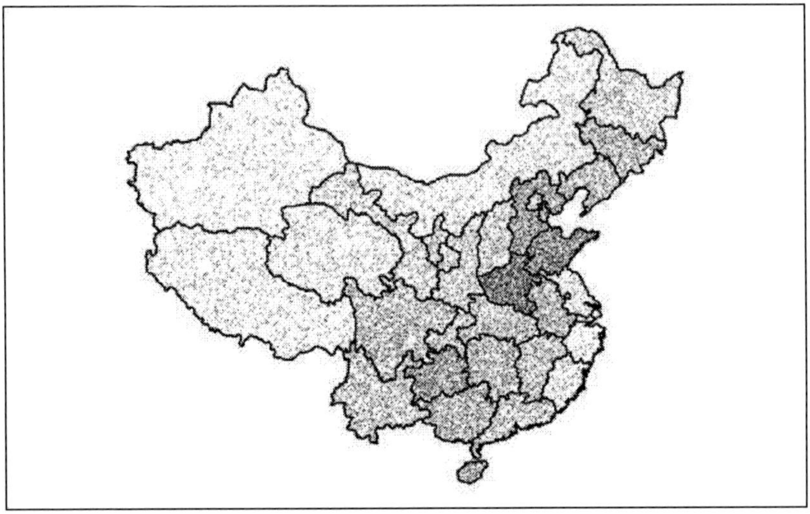

Figure 5.13: Distribution of bovines by province—2005
Source: China Livestock Yearbook, 2006. One dot represents 10,000 head

The categorisation of zones in this section follows the "beef zones" that are used in China and previous work with several exceptions.[69] Table 5.3 shows the zones (on the top row) which are ordered (from left to right) in terms of the intensity of their systems (from low productivity systems in the southeast to finishing systems in Beijing). However the zones are discussed below in order of their size and importance in the national industry. Data for these systems is shown for 2005, with average annual changes over the three years since 2003.[70] The rows show a series of production, technical[71] and income related indicators. The technical and income indicators represent

[69] Jiangxi and Hunan have developed substantial cattle production sectors so have been moved from the "other" to "Southwest" zone. Beijing and Tianjin have been moved from "other" to "finishing" systems to highlight their systems. Provinces that comprise the zones are listed in Table 5.3.

[70] The short time period is due to a lack of scale and other data for earlier periods.

[71] Rae *et al.* (2006) use official statistics to estimate total factor productivity in China's livestock sector decomposed into technical efficiency and technical change. They find that the rate of increase in productivity in beef is significant, due to technical change but not due to increased technical efficiency. Productivity growth is found to be highest in Western China and lowest in Central China.

simple averages only. Thus, the analysis does not differentiate between the significant differences between provinces within the zones. Furthermore, even within the provinces (prefecture and even county levels), there are very large differences, as examined in Brown *et al.* (2002b). Despite the aggregation in this section, some consistent patterns emerge.

5.5.1 Central Plains zone

The Central Plains zone has been the focus of the Chinese beef industry for more than two decades now. The zone built the national industry based on access to cheap low grade feed (mainly straw), draught cattle and easy access to beef markets. This "base" was developed through breed, feed and veterinary systems, the development of associations and infrastructure and a range of programs including Straw for Beef, the Favoured Areas Program and the World Bank Beef Cattle Smallholder project.

In 2005 the zone accounted for 33% of bovine numbers. Reflecting the more commercialised systems, the zone accounted for around 50% of bovine turnoff and meat production. While most production indicators have increased since 2000, the importance of the zone in the industry has declined relative to other beef cattle zones. Beef cattle numbers in most Central Plains provinces are reaching saturation point or are declining. Small unspecialised cow–calf households have begun to divert labour from cattle production into other more lucrative agricultural and off-farm activities, while the cow–calf producers left in the sector have begun to cull unproductive older and unused draught animals and turn off offspring at an earlier age. Thus, turnoff rates have increased but carcass weights have plateaued.

There are, however, major differences between provinces in the zone. In Shandong, there have been marked declines in beef cattle and breeding female numbers (as mentioned above), but between 2000 and 2005, turnoff numbers increased by 24% and beef production increased 17%. There has been a reduction in the number of unspecialised cow–calf producing households in Shandong, accompanied by an increase in specialised feeding operations. A similar pattern arises in Hebei, although the statistics are influenced by dairy cattle. All forms of cattle production are in rapid decline in Anhui, which is a correction to the over zealous expansion in the province in the 1990s.

These patterns are different to the less developed and more remote Central Plains provinces of Henan, Shaanxi and Shanxi, where there has been a sustained increase in all types of cattle production. Henan now accounts for 10% of China's bovine herd and has taken over from Shandong as the biggest livestock province in China, although in aiming for this title the provincial officials have developed a reputation for statistical over reporting. The Central Plains is the focus of analysis in Chapters 6 to 9 of this study.

5.5.2 Southwest zone

The Southwest Zone accounts for a high percentage (33%) of the bovines in China.[72] There has been a steady increase in the number of bovines in the region since 1999, led by the provinces of Sichuan, Guizhou and Hunan. The systems in the Southwest are unproductive in terms of conversion of feed to liveweight to beef. However, this form of "commercial" cattle production constitutes only one part of the functions of cattle in Southwest China. Cows are used for draught and transport functions in cropping systems, which are often hilly areas in which large framed cows are not suitable. The small cows are fed on a low grade and low cost diet which affects cycling, lactation and calving rates. Thus, calves (produced every two or three years) are also fed on a low grade and low cost diet, used as replacement draught animals or sold off at slaughter weight at a relatively old age (two or three years) rather than sold to specialised fattening households as is the case in other more specialised systems.

Household production systems are also influenced by the low levels of economic development in the region. Compared to eastern parts of China, there are fewer off-farm employment opportunities for householders, especially women (who usually look after cattle), which means they have low opportunity costs for the labour required to tend the cattle and the relatively time consuming practice of ploughing by draught power.

The predominant systems in the Southwest resemble those of the Chinese industry as a whole in the early 1990s. Cattle numbers in the Southwest have

[72] It is relevant to note that "yellow" (beef) cattle account for only 63% (with provincial variation) of these bovines in the Southwest, which includes large numbers of buffaloes. However, even taking this factor into account, the Southwest is a major beef zone. The buffalo meat enters the mass bovine meat market along side and is often undifferentiated from meat from the yellow cattle. The buffaloes also impact on the technical indicators of turnoff rate and carcass weights. However, these indicators are similarly low for the yellow cattle herd. Furthermore, the indicators of female numbers and scale data are for yellow cattle only.

grown faster than they have in the Central Plains over the seven years since 1999. The Southwest is likely to follow the Central Plains and Northeast into developing more productive systems. There are already many "pockets" that are entering into commercialised household cattle operations such as Ya'an Prefecture in Sichuan and parts of Jiangxi and Hunan. Furthermore, there are prospects that the Southwest may fill some of the diminished supply created by the exit of smallholder cow–calf and small feeding operations in the Central Plains and the Northeast. This however, depends on the viability of moving feeder cattle in the Southwest to finishing operations in the Central Plains, which is closer to more developed final beef markets. However, it is also possible in the longer term future that more cattle will be finished and slaughtered in the Southwest and the beef supplied to both local and inter-provincial markets, if the infrastructure and especially cold storage facilities are upgraded. Even in this case, however, Central Plains provinces like Henan and Liaoning would seem to hold a comparative advantage in cow–calf production into the medium term future.

5.5.3 Western pastoral zone

The Western Pastoral zone has traditionally been associated with ruminant livestock production before the focus of the industry shifted toward the Central Plains in the 1980s. The zone still contains a significant proportion of China's bovines (19% in 2005) and has increased in relative importance since 1999. While this increase is largely due to the increase in dairy cattle in Inner Mongolia and Xinjiang[73], beef cattle numbers in Ningxia and Gansu have increased substantially. Another factor that makes it difficult to generalise about developments in the Western Pastoral zone is the heterogeneity of the production systems. There are few cattle in "pure" extensive pastoral systems and even these pastoral areas have undergone a policy induced intensification of production systems in recent years (Brown *et al.*, 2008). Most of the beef cattle are located in agricultural systems (which are not dissimilar to agricultural parts of China) and semi-pastoral systems (with some access to grazing land). The semi-pastoral systems mean that the scale of beef cattle production is relatively high. This production system is based on cow–calf systems with households keeping on calves and feeders for highly variable time periods.

---

73 Dairy cattle make up nearly 50% of the bovines in Inner Mongolia, while yaks dominate bovine numbers in Tibet.

Table 5.3: Key production, technical and revenue indicators for six regions

Production indicators	*China*		*Southwest (a)*		*Other (b)*		*Western (c)*		*Central Plains (d)*		*Northeast (e)*		*Finishing (f)*	
	2005	**Av. an. increase since 2003 (%)**	**2005**	**Av. an. increase since 2003 (%)**	**2005**	**Av. an. increase since 2003 (%)**	**2005**	**Av. an. increase since 2003 (%)**	**2005**	**Av. an. increase since 2003 (%)**	**2005**	**Av. an. increase since 2003 (%)**	**2005**	**Av. an. increase since 2003 (%)**
Proportion of China's bovine herd in region	100		33		5		19		33		10		0	
Bovines in region ('000 head)	141,575	1.7	46,269	1.4	6,620	–2.3	26,944	6.7	46,838	0.2	14,218	2.7	688	–2.4
Bovine turnoff in region ('000 head)	52,876	4.1	10,836	7.7	1,395	6.5	8,258	6.9	23,348	1.9	8,417	4.2	623	–0.4
Bovine meat production from region ('000t.)	7,115	4.3	1,138	7.5	146	3.4	1,060	7.9	3,422	1.8	1,243	6.1	106	–0.1
Technical indicators														
Av. prop. Productive females in cattle herd (%)	44	0.1	44	1.7	37	0.3	42	-1.9	45	0.1	56	2.9	17	–3.0
Av. bovine turnoff rate (%)	37	0.8	25	1.2	20	1.0	32	0.1	44	0.7	62	0.8	93	1.5
Av. bovine carcass weight (kg/head)	135	0.2	106	–0.3	156	–10.8	123	1.0	148	–0.2	148	2.2	172	0.7
Av. cattle turned off in 1–9 head scale cat. (%)	65	–2.4	86	–1.1	80	2.9	58	–3.5	77	–0.9	43	–1.0	10	0.1

Av. cattle turned off in 10–49 head scale cat. (%)	19	1.1	10	0.7	12	2.1	28	2.8	14	0.2	32	2.6	44	4.0
Income indicators														
Av. per capita rural revenues (Rmb)	3,255	8.0	2,610	7.8	5,411	6.9	2,365	7.5	3,280	8.0	3,392	9.2	6,463	9.0
Av. proportion revenue earned on-farm (%) (g)	61	0.5	60	0.6	48	–0.8	73	0.7	58	0.2	76	–0.2	44	1.1
Av. proportion revenue earned from livestock (%)	19	0.4	25	0.6	10	0.2	25	0.2	16	0.3	22	0.6	12	0.1
Av. proportion revenue earned off-farm (%) (h)	39	–0.5	40	–0.6	52	0.8	27	–0.7	42	–0.2	24	0.2	56	–1.1

Source: Derived from China Livestock Yearbook, China Statistical Yearbook, and China Rural Statistical Report, 2004 and 2006
(a) The 7 "Southwest" provinces are Guizhou, Sichuan, Guangxi, Yunnan, Jiangxi, Hunan and Chongqing
(b) The 5 "Other" provinces are Shanghai, Guangdong, Hainan, Zhejiang and Fujian
(c) The 6 "Western" provinces are Tibet, Qinghai, Gansu, Inner Mongolia, Ningxia and Xinjiang
(d) The 8 "Central Plains" systems provinces are Hubei, Henan, Anhui, Shandong, Jiangsu, Shanxi, Hebei and Shaanxi
(e) The 3 "Northeast" provinces are Jilin, Liaoning and Heilongjiang
(f) The 2 "Finishing" systems are Tianjin and Beijing
(g) Data on the proportions of revenue from various activities are sourced from National Bureau of Statistics reports that cover the first three quarters of 2003 and 2005 only.
(h) Revenues earned off-farm include wages, earning on assets (such as land) and transfer payments (including from part time labour)
Note female proportions and scale data refer to "yellow cattle", while turnoff numbers and carcass weight refer to bovines.

Statistics indicate that the increase in bovine numbers has not been accompanied by an increase in productivity. This is partly due to the influence of dairy cattle. However, even without this factor, it is widely acknowledged that beef cattle production systems in semi-pastoral areas are inefficient. Feeder cattle are kept in the household system for long periods and undergo a cycle of weight loss in winter/spring (when on winter feed in pens) and weight gain in summer/autumn (when grazed on pastures). Most cattle are sold off after this grazing period but there is scope to increase turnoff rates in more efficient systems. There are also ample sales channels for young grass fed feeder cattle to other areas. These semi-pastoral systems are not addressed directly this study but are discussed in (Longworth and Williamson (1993) and Brown *et al.* (2008). Waldron, Brown and Zhang (2007, Appendix 1.1) examine the Inner Mongolian beef cattle system in detail.

5.5.4 Northeast zone

The Northeast accounts for 10% of the bovines in China and the numbers and relative importance is increasing. This is significant given that a large proportion of cattle in the Northeast are improved breeds, that the turnoff rates and carcass weights are relatively high and that the scale of production is relatively high. The increases in cattle numbers in Liaoning occurred uniformly over the small and medium scale categories, while numbers in larger feedlots declined. There appear to be anomalies in the scale data in Jilin, which indicates a major exodus out of small-holder cow–calf systems, accompanied by an enormous move into feeding in the mid sized fattening household and feedlots, but not in the larger feedlot categories.

It appears that the cattle industry in the Northeast is beginning to realise its comparative advantage in cattle feeding, with an abundance of crop (especially corn) residues and grain (and lower grain prices). There are also large numbers of slaughter units and abattoirs in Jilin and large numbers of cattle dealers and markets transport an increasing number of feeder cattle to provinces like Shandong, Hebei, Beijing and Tianjin for finishing. However, this is also backed by a strong cow–calf production system in provinces like Liaoning, where there are still undeveloped areas like Tieling Prefecture.

5.5.5 Urbanised Beijing-Tianjin zone

This is zone is unique in that it is based on cattle finishing and, although small, is overviewed briefly here to highlight the extreme of the Chinese cattle and beef production system. In particular, it provides insights into the urbanisation of livestock production. Most slaughter units (which are usually designated slaughter points in these cities) slaughter in early hours of the morning to get the fresh beef to market by early morning. The slaughter units therefore draw on a flow of finished cattle from specialised fattening households, the majority of which are still located within the city boundaries, although this is expected to decrease over time. There are also a number of integrated feedlots and abattoirs in these cities. City officials encourage primary production activities within the cities to increase food security and reduce risks of disease in the event of an outbreak like Sudden Acute Respiratory Syndrome or Avian Influenza, which can block inter-provincial transport of live animals. For a more detailed overview of beef and sheep meat production and marketing in Beijing see Brown *et al.* (2002b, Appendix 15) and Waldron, Brown and Zhang (2007).

Both the data and fieldwork suggests that an increasing number of stallholders in wholesale markets are switching sources from within the cities to other nearby areas outside the city, including Hebei (Sanhe and Dachang) and Shandong (Yangxin and Qinghe). This requires transport of beef over significant distances, which must in law be refrigerated to enter the Beijing City. Cattle and beef production systems in these areas, which are examined in detail in Waldron, Brown and Zhang (2007), are similar to those suggested in the data for Beijing and Tianjin.

5.5.6 "Other" zone

This zone is very small, accounting for 5% of China's bovines and is comprised of wide range of provinces such as Fujian and Shanghai. It is also diverse, making generalisation about systems problematic. However, the "other" zone still contains some substantial operations (feeding, finishing, slaughter) for specific east coast markets. Given access to off-farm opportunities, there appears to be limited incentives for east coast households to engage in cow–calf production, as reflected in diminishing bovine numbers.

5.6 Concluding Comments

A careful interpretation of official statistics provide indications of the way that the Chinese beef industry has developed over the post-reform era and how it might develop into the future. The growth of the beef industry has been led partially by an increase in beef consumption generated by increasing incomes, urbanisation and out-of-home consumption. However, the consumption of beef in general has increased at only a modest rate while the consumption of premium beef in particular remains very low. These market settings need to be better understood and incorporated into the decision-making rationales of policy makers and investors, who tend to overstate the potential of the beef market based on the experience of countries other than China itself.

Rapid industry growth in the 1990s was also driven by aggressive and effective policies to establish a beef cattle "production base". Policy emphasis then switched to other industries, beef demand remained subdued and household producers diverted resources into more lucrative off-farm activities, which led to a plateau and then decline in beef cattle numbers in the 2000s. However, more of this cattle herd is now oriented to "commercial" beef production rather than for draught power. The extent to which this translates into beef output is limited by the incentives of producers to turn off cattle at a young age and at low carcass weights. Further rationalisation of the industry can be expected, especially amongst unspecialised households in the relatively developed Central Plains area of China, and amongst large feedlots. Specialised household producers in slightly less developed areas are, however, are becoming increasingly prominent in the beef cattle production sector. How this evolving production sector gains from industry participation depends on the development of markets and other sectors of the industry, as discussed in detail in the following chapters.

CHAPTER SIX

DEVELOPMENTS IN CONSUMPTION, RETAILING AND PRICES

While a circumspect analysis of official statistics provides some insights into developments in the beef industry, the insights are limited by the degree of aggregation and the inaccuracies in the statistics. As a result, the following Chapters 6 to 9 adopt a more disaggregated approach by drawing on fieldwork and budgeting data in specific sectors of the industry.

The first of these chapters examines beef consumption, prices and retailing on the basis that these downstream parts of the industry "pull through" upstream parts of the industry examined in subsequent chapters. The impact that downstream aspects of the industry have on upstream parts of the industry, therefore, must be well understood in order to assess industry development options. However, prices and consumption are also impacted by production aspects of the industry—product that can be viably supplied to the market—in ways also discussed in this chapter.

The chapter overviews consumer preferences and the product attributes demanded in the market (Section 6.1), presents price information in different retail outlets (Section 6.2) and beef prices ex-abattoir (Section 6.3). This price and product information is used to identify three beef value markets—low, mid and high—as presented in Figure 6.1, which is an extract from the main study framework (Figure 3.1). Further information on consumption, retail channels and prices is presented in Table 6.1, Table 6.2 and Table 6.3. The price and product information which form inputs into the retail and consumption sectors, also form the outputs of the next upstream stage of cattle slaughtering which is analysed in Chapter 7.

The analysis in the chapter draws on other studies that find that beef (like other meats) as a broad product class is a normal good in so far as consumption increases with incomes. However, within the broad product class of "beef", there is little evidence to support the widely held impression amongst policy makers and new market entrants that China will follow Japan, South Korea or many Western countries in the development of a large and growing market for high value, premium beef with exacting quality attributes. The chapter argues that while a "modern retailing sector" (especially supermarkets) are selling more beef, the share of the overall market is insignificant, and only a small proportion of the beef sold through supermarkets is "high quality", purchased through "modern supply chains" or hold any implications for cattle producers or traders. However the analysis argues that there is a growing market for mid value beef that is sold through the full range of outlets—from traditional markets to new restaurant franchises—in which food safety and hygiene is the most important attribute demanded. These basic market characteristics are very often over-looked by policy makers, investors and in the literature—both within and outside China—with some major detrimental implications, as established in subsequent chapters.

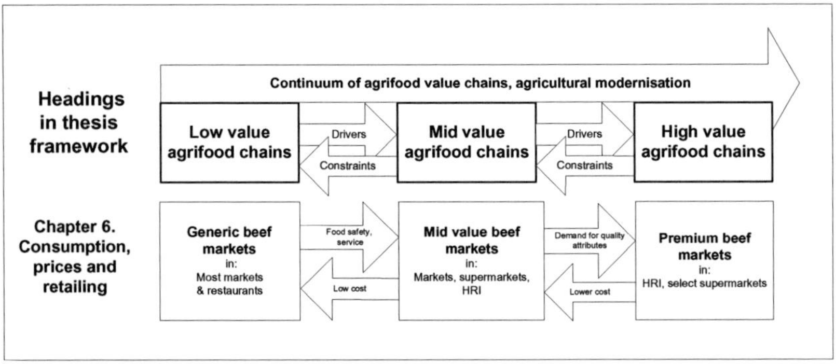

Figure 6.1: Outline of structures, drivers and constraints to modernisation of the consumption and retailing sectors (extracted from Figure 3.1)

6.1 Consumer preferences and product attributes

The review of official data in Section 5.2 provided an overview of broad changes in beef consumption. But an appreciation of where beef consumption in China is heading requires an understanding of the drivers to that change based on a more disaggregated, detailed and comprehensive information than what the official statistics can provide. This section examines the consumption trends through more detailed survey and fieldwork information. Consumer preferences are related to specific product attributes demanded which, as shown below, vary enormously between market segments.

Beef consumption levels in low value markets are gradually increasing due to perceptions of its nutritional value (protein), health benefits (leanness) and as a result of the diversification of traditional diets. There appears to be significant growth and growth potential in the mid value market, where a growing number of increasingly affluent consumers are prepared to pay for safety assured product along with basic quality attributes for specific dishes consumed out-of-home. There is little evidence that the small market for high value, premium beef will grow significantly. Despite the importance of consumption trends to industry development, they are not well understood in China or reflected in industry policies and investment decisions. A summary of these points appears in Table 6.1 and are discussed in more detail below.

Table 6.1: Size of and product attributes in low, mid and high value beef market segments

Row		Low value market segment	Mid value market segment	High value market segment
1	Estimated size of market (2006, in volume, across China)	• Dominant, 80–85%	• Significant, 13–15%	• Small, 2–5%
2	Product attributes	• Generic beef, with few price differentials between cuts • Lean beef • Basic State hygiene inspection, not always enforced	• Particular cuts, with some quality attributes • Food safety certified through specific schemes	• High value, large cuts • Particular quality characteristics • Safety assumed
3	Growth potential (in volume terms in urban areas)	• Modest in-home consumption • Higher out-of-home	• Significant out-of-home • Significant in-home consumption	• Negligible in-home consumption • Modest out-of-home consumption

6.1.1 Consumer preferences and product attributes in the low value market

Generic beef sold in the low value or mass market accounts for between 80% and 85% of all beef sold in China. Consumers in low value markets have very low expectations and requirements for quality characteristics of generic beef but are becoming more selective about the health and food safety attributes. Consumers generally have a positive impression about the health and food safety attributes of beef, which may provide a source of some price premiums into the future, if the supply chains are managed and regulated effectively.

Health and food safety attributes

The most persistent theme in surveys of the generic beef market is that Chinese consumers are becoming more health conscious and are seeking to increase protein intake through increased meat consumption. At the same time, in contrast to the past when "more fat was better",[74] they are now seeking to minimise animal fat intake, as found in various studies[75] and on fieldwork.[76]

Beef is perceived as a lean meat in China because of supply-side cattle production practices. Chinese cattle breeds have been bred for work (draught, transport) rather than rapid weight gain and are fed on cheap, low grade feed, especially crop residues. The beef from these cattle is much leaner than grain fed beef or Chinese pork. Given the concurrence between consumer preferences in the mass market and cost structures in cattle production, Chinese beef is likely to continue to remain lean at least in the low value mass market.[77]

Longworth *et al.* (2001, Chapter 13) found that by far the most important characteristics sought by consumers in beef is freshness and cleanliness

---

[74] This is for taste reasons and also because fat from the meat was commonly used in the past to grease the pan for cooking, whereas cooking oil is used more now.

[75] Longworth *et al.* (2001, Chapter 13) found that Chinese consumers in the mass market in regional cities prefer beef with no fat at all, considered even a little fat around the edges of the meat to be undesirable, and place no or little value on marbling. Meng (2007) found that consumers' concern about high blood pressure increases the demand for beef, sheep and goat meat, egg and dairy and decrease the demand for pork. Wang *et al.* (2004) find that the vast majority of beef (87%) is consumed as lean meat—and relatively little as by-products, ribs or fatty meat—far more than for pork and sheep meat.

[76] Stallholders in markets throughout China say that Han Chinese customers do not want fatty beef or sheep and goat meat (more than 4–5mm external fat and excess seam fat) and will ask the stallholders to slice off the fat. This reduces the weight and therefore the margins of stallholders, especially as the fat is sold at a heavily discounted price of about Rmb4/kg. However, some minorities such as Hui and Mongolians prefer fat beef (and sheep and goat meat).

[77] The differences between beef produced in China and Western countries is not taken into account by the President of the Physicians' Committee for Responsible Medicine who is critical of China's efforts to build a beef industry on health grounds and that "the World Bank's efforts to help China make a killing in cattle farming is aimed not at good health or a clean environment, but at economic investment" (Barnard, 1999). Others such as McMichael (2001) argue on a global level that the growth of the beef industry is dependent on medicines, antibiotics, chemical fertilisers and herbicides, and that beef is harmful to health because it displaces cholesterol free vegetarian based diets in developing countries. Again these broad statements do not apply to the bulk of beef in China.

(hygiene), followed by the related attributes of smell and colour. Meat in the mass market is sold fresh on the day of slaughter and then transported to market. Veeck and Burns (2005) found that consumers prefer to buy meat that has been slaughtered that day and has not been refrigerated (at processor, retail or consumer levels). In the late 1990s it was very common to see beef that had been injected with water to make it appear more juicy and also to increase the weight of the cut.[78] This practice has been banned (by the Ministry of Health) and while fieldwork suggests that this has been partially effective, it is still common, even for beef produced in large abattoirs. In wet markets without refrigeration, meat can be distinctly "unfresh" by the end of the day, and conditions at slaughter, distribution and retail are highly variable. Given these realities, consumers reduce safety hazards at point of purchase by their own sensory appraisal and by changing purchasing practices.

The major and highly publicised problems with food safety in China in recent years have, however, meant that the State is turning attention toward food safety problems. This is reflected in a gradual introduction of basic inspection at pre-slaughter and post-slaughter stages in more developed parts of China.[79] Stallholder facilities are also routinely inspected by local Health Bureaus in these areas. Heavy fines are notionally imposed at the vendor level.[80] There is of course a large gap between policy and practice, especially in small wet markets.

Even in the mass market, however, Chinese consumers have a positive perception of the food safety attributes of beef compared to other meats, again due to production side practices. Consumers in China are becoming aware—through the media and word of mouth—of the hormones, antibiotics and by-products fed to livestock that flow through to the end product. These sort of intensive production practices are more closely associated with the

[78] This accords with some of the practices reported in Veeck and Burns (2005) where consumers were wary of meat that is too red and where vegetables had been soaked in water.

[79] In Beijing and regional cities, beef at markets is generally accompanied by three forms of certification issued by local Animal and Quarantine and Inspection Departments: one for animal disease inspection as the animal enters slaughter point; one for carcass inspection post slaughter; and an edible dye stamp on the carcass to clear the meat to be sent to markets.

[80] In the case of Gansu, according to Anonymous (2005), meat from livestock slaughtered in private slaughter households is not allowed to be sold at market. Neither is meat that has been injected with water. If caught, vendors can be fined two to five times the value of their trade and can have their licenses withdrawn. Slaughter units that sell meat that does not meet national health standards can be fined between Rmb3,000 and Rmb50,000. All meat must be inspected and show evidence of inspection or there can be fines of Rmb2,000 to Rmb30,000.

more intensive production systems of pigs, poultry and fish than with the less intensive, "greener" and smaller scale cattle production systems. In addition, unlike pigs, poultry and dairy, beef cattle have not been associated with any widely reported disease or contamination outbreaks in China. On the other hand however, as discussed in the Chapter 7, slaughter practices for beef (and sheep and goat meat) are far less regulated than for other meat. As is the case of other foods, China has developed a large number of food safety standards and regulations for beef.[81]

The generally positive perception of beef as a safe meat provides opportunities for industry development. However, this requires a series of measures to improve consumer perceptions of food safety of beef, including the enforcement of food safety regulations, the effective management of slaughter points and increasing trust in public certification schemes. These schemes will pose additional costs and will impact on the livelihoods of uncertified slaughter households when they are banned. However, it is argued in Chapters 7 and 10 that the promotion of unmechanised certified slaughter points is an appropriate response because they will facilitate the improvement of food safety standards, have low cost structures and will be able to absorb a large numbers of uncertified slaughter households and workers when they are effectively banned from operating on an individualistic basis.

---

[81] The "Quality Agricultural Products Development Service Centre of the Ministry of Agriculture" issues a range of standards, including for beef, called the "Safe and Hygienic Quality Beef Standards". These standards are comprised of two "subsidiary" standards: the "Safe and Hygienic Quality Beef Production Technology Operational Model"; and the "Safe and Hygienic Quality Beef". The standards cover the full range of feed and veterinary products in production and that can act as residues. Inspection must not reveal cases of pathogenic bacteria, Foot and Mouth Disease, certain parasites, Bovine Spongiform Encephalopathy, Clenbuterol Hydrochloride suppositories, Colchicine or Diethylstibestrol. It is claimed by various industry participants that the growth promotant Clenbuterol Hydrochloride is commonly used in China to increase muscle mass gain. In addition, the standard for "*Wugonghai—niurou*" ("Non Public Hazard—Beef") was issued on September 3, 2001 by the Ministry of Agriculture and enacted on October 1, 2001. The standard prescribes the definition, technical requirements, inspection methods, symbol, packaging, stockpile and transportation of *wugonghai* beef. It makes reference to a large number of national hygiene and measurement standards (GB and NY codes) that are general across the food and meat sectors. Other programs related specifically to cattle include the "*wugonghai* foods beef cattle raising veterinary products usage standards" (NY 5125), the "*wugonghai* foods beef cattle raising veterinary products disease prevention standards" (NY 5126), the "*wugonghai* foods beef cattle raising feed use standards" (NY/T 5127), and the "*wugonghai* foods beef cattle raising management standards" (NY/T 5128).

Other quality attributes

Surveys suggest that tenderness and juiciness are the least valued quality attributes by consumers in the mass beef market in contrast to Western countries and Japan where they are highly valued. This is partly because tenderness and juiciness are not required in the vast majority of Chinese dishes that use beef which are, in order of importance, stewing, five spice roasting and stir frying (Longworth *et al.*, 2001, Chapter 13).

The partiality of Chinese consumers to tough and dry beef relates to the type of product that they have traditionally been supplied with. Well into the 1990s, much of the beef offered on the mass market was from culled draught cows that produce tough, dry, dark meat with yellow fat. However, the general "quality" of Chinese beef has improved over recent years due to breed improvement, higher turnoff rates and lower carcass weights discussed in Section 5.4 above. As also discussed in Sections 9.1 and 9.2, household cattle producers have economic incentives to turn off feeder and slaughter cattle at an increasingly young age in order to reduce feed and other costs, which produces beef that is more tender, juicy, light in colour, with whiter fat but with less taste. The widespread availability of this type of beef on the mass market might mean that a wide range of consumers develop a greater appreciation for different types of beef and beef quality attributes.

For markets where fat is required—hamburgers, some of the processing trade and for hot pot and beef skewers—lean Chinese beef often has to be supplemented with cheap animal fat purchased (and imported) as a separate product. The fat is mixed with beef or, in the case of hot pot meat, infused into the meat rolls. While there are many instances of "fake marbling" (through fat infusion) for steaks and Japanese barbeque, "typical" generic Chinese beef is not suited to these higher value markets because it is too lean, unmarbled, can only be butchered into small prime cuts, and is not "tasty". Production of beef to suit these higher value markets is supplied through completely different supply chains to that for generic beef, as discussed below.

6.1.2 Consumer preferences and product attributes in the mid and high value markets

While surveys of consumers in the mass market provide insights into consumer preferences for beef in general, they do not capture consumer preferences and product attributes for beef in mid and high value markets.

Data on this is not widely available so the discussion draws on fieldwork visits to retail and service outlets.

Health and food safety attributes

With regard to mid value markets for in-home consumption, consumers are prepared to pay modest premiums for safe foods (Wang, 2003; Wang, 2006; Yang, 2005; Wang *et al.*, 2008). Food safety price premiums are most relevant to certification schemes such as the "Green Foods" (*luse shipin*), "Non Public Hazard Foods" (*wugonghai shipin*) and "Organic Foods" (*youji shipin*) programs of the Ministry of Agriculture and the "Rest Easy Meat" (*fangxingrou*) program established under the Vegetable Basket Program. It is claimed that premiums of 30% are possible for Non Public Hazard Foods and Green Foods certified product (Ministry of Agriculture Green Food Office, personal communication), although this is certainly an upper limit.

While food safety attributes are in demand and offer potential to increase premiums, problems have emerged in practice. In all cases, certification schemes require close monitoring back to production level within tightly integrated value chains. This applies significant transaction costs on the value chain that must be offset by significant premiums. Fieldwork suggests that premiums for most foods, including beef, are not as high as reported above by the administrators of the schemes and in most cases are not sufficient to offset additional costs. In addition, it is widely believed in China that the schemes have not been sufficiently monitored and governed so that consumers have lost trust in the certification schemes. This further reduces the premiums that can be generated through the schemes.

The lack of trust in public systems is one of the reasons why large companies use their reputations and brand names to generate trust and to differentiate product to generate premiums. This is reflected not only in the form of company branded products, but also in company (agent or franchise) stalls at markets and company counters at supermarkets, both of which are very common. "Branded agriculture" (*pinpai nongye*) has become a major movement in recent years that is formally and officially encouraged by the Ministry of Agriculture.

Consumers in both large and regional cities also increasingly rely on retail outlets as a means of providing more certainty in food safety. Most attention is usually paid to the supermarket sector, which is generally more trusted than wet markets to provide safe food, including meat. In many regional supermarkets, the premiums are not necessarily—and indeed are not often—derived from the attributes of the product *per se*. The same type of meat sourced from a given slaughter unit will sell for a higher price in a supermarket than in a market. In this regard, supermarkets generate retail premiums from the facilities (cold storage, air conditioning, glass cabinets, stainless steel cutting boards etc.), presentation, reputation and convenience of the outlets and the certainty that the supermarkets have sourced product from centralised certified slaughter facilities rather than uncertified slaughter households.

Quality attributes

There is also evidence that consumers in the mid value beef market are beginning to develop preferences for quality characteristics such as tenderness, juiciness, colour, fat and marbling. This is expressed in Table 6.2 (Rows 5 to 13) where small price premiums for particular cuts in the low value markets become large price premiums for particular cuts in high value markets where quality attributes become increasingly important. For example, restaurant dishes like hot pot and American fattened beef require significant external (sometimes intra-muscular) fat, while Japanese barbeque requires marbling, and Brazilian barbeque requires large primal cuts of lean beef. Required quality attributes must be met through specific supply chains and industry participants in the slaughter, marketing and production sectors.

While there is significant anecdotal evidence to suggest a significant and growing mid value market, there is little evidence of a significant high value (premium) beef market. The domestic high value market is confined to the large cities of China and mainly to the Japanese, Korean and western expatriate and tourist market. This market is affluent and demanding of beef characteristics similar to those available in home markets. However, the size of this market has been vastly over estimated by policy makers.[82] The over estimation means that the high value market is over supplied by a large number of feedlots and abattoirs as discussed in Chapter 7.

[82] This mindset is predominant across all levels of government and industry. At the highest level, a cattle nutritionist who is head of the China Cattle Association and the major industry advisor to government, claimed that annual per capita beef consumption in Beijing was 20kg, that much of this was "quality beef" consumed out of home, that there was no need to worry about surplus capacity in the abattoir sector because Chinese beef consumption will follow international patterns (China–Australia Free Trade Agreement technical workshop, Beijing, July 2006).

Furthermore, high value export markets are likely to remain inaccessible at least into the medium term due to disease protocols. The constraints to development of a high value beef market need to be better reflected in more feasible policies and investment in the slaughter, marketing and production sectors of the industry.

Beef Quality Standards

Quality standards (or grading systems) are a major aspect of the development of mid and high value markets. In order to increase consumption and increase price-grade differentials, the industry notionally has an interest in providing customers and consumers with a product that consistently meets more specific requirements. Sale by description—rather than through subjective assessment and personal relationships—also becomes an increasingly efficient way to trade as the volume and distances of trading, and the number of buyers and sellers in the market increases (Li and Li, 2000). In recognition of these demands, China has developed several beef grading systems, the most important of which is titled "Beef quality grades" of the Ministry of Agriculture.[83] The standards are presented and analysed in full in Waldron, Brown and Zhang (2007, Section 3.4.8 and Appendix 2).

Despite the apparent need for—and the formal existence of—adequate quality standards, uptake has been low and slow for several reasons. First, quality standards are not relevant in the mass market where there is little demand for information about quality differentials. Second, in the abattoirs active in high value market seek to differentiate their product based on their company reputation ("branded agriculture") rather than by competing directly with other abattoirs based on common quality standards. Third, adherence to quality standards reduces room for opportunism, both in the way that abattoirs sell beef outputs and purchase cattle inputs. Fourth, there is adversarial competition and replication between institutions that develop and administer beef quality standards, including between ministries and research institutions.

[83] "Beef quality grades" (NY/T 676-2003) issued on 30th July 2003 by the Ministry of Agriculture and enacted on October 1, 2003. Other standards include: The Ministry of Commerce/Internal Trade slaughter standards mainly for the General Food Company system; the standard "Frozen Beef" issued in 29th March 2004 by the General Administration for Quality Supervision, Inspection and Quarantine; the Shandong Department of Science and Technology has developed beef standards; and the other food safety schemes listed in Section 6.1.1 are also related to quality aspects of beef marketing.

Fifth, any single set of quality standards in China can not cater for the varied (Western, Korean, Japanese, South American and Chinese) styles of cooking, customers and consumers, all of which share a roughly equal place in the high value Chinese beef market. That is, the small premium (or high value) market in China comprises of a series of even smaller segments each with distinct requirements and features. In this regard, it is relevant to note that "Beef quality grades" (NY/T 676–2003) is a grading system that applies a single grade (e.g. A, B or C) to a given product (order of meat). Because a single grade may not be a useful reflection of the value of the meat to different customers (i.e. a large amount of information about the product is "lost" in a single grade), it may be more effective to use a classification system that provides a greater number of measurements on a number of attributes which then allows customers or consumers to better choose the products that suits their requirements. Even in this case, however, developing quality standards that cater for a fragmented market in a relatively inexpensive, logistically feasible way is a very challenging task in China.

As a result of the lack of uptake of beef quality standards, beef marketing in the high value market is described by industry experts as "chaotic" (*luan*). Beef grades based on in house grading systems (to produce and market product labelled as "high grade beef" (*gaodeng niurou*) for example) are not comparable across abattoirs, thus hindering competition and precision in trading. There can be differences in the way that cuts are butchered (to include or exclude particular muscle areas). Even the names used to describe a particular cut very often vary between regions and abattoirs. Buyers—such as purchasing departments of hotels, chefs of restaurants or general commodity traders—without a long history in the beef trade or who are not familiar with locally used terms and slang can easily be confused by the multitude of terms. Thus, an important initial step in standardising trade is to develop a common "beef language" in China.

The lack of "standardisation" in the beef industry entails a series of implications. The inability of customers to compare products and prices across suppliers (abattoirs or beef traders) reduces competition in the industry and increases product risks that are compensated by price averaging. Customers that can not afford to take product risks (such as five star hotels) turn to overseas sources of supply. Uncertainty about the product amongst customers and consumers also affects longer term confidence in the product and constrains beef consumption and the development of price-grade differentials. The lack of subjectivity in, or inconsistency between, quality standards also affects the way in which cattle are purchased because beef quality standards also frame the standards by which live cattle are purchased, as discussed further in Chapters 8 and 9.

While there are some significant difficulties in the adoption of a widely used quality standards, especially for the high value market, it is clear that the "unstandardised" status quo acts as a constraint to the development of beef markets, with negative implications for industry participants in rural areas. As discussed further in Chapter 10, quality standards may be more relevant to mid value markets and mid sized industry actors that do not have a strong company name or brand name on which to differentiate their products in the market place.

6.2 Prices and characteristics of various retail channels

Official beef price series were presented in Section 5.2. However, this data is limited in that it relates only to generic beef. Other price information collected on fieldwork allows for a more disaggregated analysis of prices for various forms of product. In Table 6.2 this price information is sorted into low, mid and high value market segments. As shown in Row 1 of Table 6.2 it is not possible to categorically sort types of retail outlets into specific market segments. Types of retail outlets—including wholesale markets, supermarkets, and restaurants—span market segments, an issue that is sometimes over-looked in the literature (see Section 2.2.2 above).

Because the "case study" markets, supermarkets and restaurants shown in Table 6.2 range from low to high value retailers, they can be seen as covering a broad spectrum of the retailing sector, although it has not been possible quantify the representativeness of each particular retailer listed in Table 6.2. Together, the information combined in Table 6.2 provides some indication of the retail level price differentials associated with market segmentation and the development of "high value agriculture". This is condensed into Row 2 of Table 6.2 to provide a guide to the price range at retail level in low, mid and high value market segments.

6.2.1 Prices in wet markets

The prices for generic beef presented in the official data in Section 5.2 is similar to those reported through a separate wholesale market reporting system, both of which are considered part of the mass market.[84] However, the price data from the wholesale market reporting system provides additional information, especially on the price differentials between different cuts. Row 3 of Table 6.2 provides an example of the type of information available in the case of Xinfadi Market in Beijing in September 2006.[85]

Table 6.2: Price information for various retail outlets in low, mid and high market segments

		Low value market segment	**Mid value market segment**	**High value market segment**
Retailing				
1	Relevant retail outlets	• Generic stalls in wet markets • Generic product in supermarkets • Low value restaurants	• Company/specific stalls at wet markets • Company counters and branded product at supermarkets • Mid value restaurants	• High value restaurants • Branded product in particular supermarkets
Prices				
2	Price range at retail level	Rmb15–25	Rmb25–50	>Rmb50
Markets				
3	Xinfadi market, Beijing, reported prices on agri.gov.cn 25 September 2006	Normal stall: beef ribs (Rmb13.6), front quarter (Rmb14.6), fattened beef (Rmb15), chuck roll (Rmb15), connective tissue (Rmb16), Rump (Rmb15.6), front shank/shin (Rmb16), quality fattened beef (Rmb16), hind shank (Rmb16), thin flank (Rmb16), loin (Rmb18), quality tenderloin (Rmb20)		—

[84] The price information for generic beef in 23 major wholesale markets across China is reported on a daily basis through the China Agricultural Markets Association—http://www.chama.org.cn/. Other price data from a wide range of wholesale markets throughout China is available on http://www.agri.gov.cn/jghq/ly/ and http://www.agri.gov.cn/pfsc/

[85] Data provided by the Xinfadi market in 2006 shows that price levels changed very little over the course of that year. Volumes of beef increase at Chinese New Year but there was little price effect.

		Low value market segment	Mid value market segment	High value market segment
4	Xinfadi market, Beijing, actual prices,19 July 2006	One normal stall sold all cuts for Rmb17/kg, except for fillet which sold for Rmb18–20/kg. The stallholder purchased half carcasses for Rmb14/kg. Another normal stallholder sold rump for Rmb16/kg, fillets for Rmb22 and other cuts for Rmb15. The stallholder purchased half carcasses for Rmb14.8/kg. A hot pot stall sold sliced frozen hot pot beef: Grade 3 (Rmb18/kg), Ungraded (Rmb15/kg).	A hot pot stall sold sliced frozen hot pot beef: grade 1 (Rmb36/kg); grade 2 (Rmb28/kg);	—
5	Ji'nan Minority Market, Shandong, July 2006	Normal stall: generic beef (Rmb17), mince (Rmb17), fillet (Rmb20), neck (Rmb16), fattened beef slices for hot pot (Rmb24), slices from hindquarter (Rmb20).	Normal stall: striploin (Rmb28). There were also agents stalls for Shuanghui, Yurun, Jinluo and Caoyuan Xingfa companies, which sell in the mid value market price range	—
6	Altay wet market, Xinjiang, June 2006	Normal stall: generic bone-out beef (Rmb18), hindquarter (Rmb17), bone/ribs (Rmb4), fat (Rmb4–9).	—	—
7	Daxi Market, Shenyang, 1998	Nornal stall: rump (Rmb15), striploin (Rmb14), breast (Rmb13)	Fucheng agent stall in market: Rib eye fillet (Rmb80), tenderloin 1A (Rmb70), tenderloin 1B (Rmb50), fattened beef No.1 (Rmb50), fattened beef No.2 (Rmb40), fattened beef (Rmb32)	—

		Low value market segment	Mid value market segment	High value market segment
Supermarkets				
8	Hualian supermarket in Ji'nan	One section of the cabinet sold generic beef from a designated slaughter point for Rmb22 for rump. Another section sold beef from the Hengdian abattoir at the prices: ribs (Rmb20.6/kg),hindquarter (Rmb23.6–25/kg), tail (Rmb26.4/kg) and beef mince (17.6/kg).	—	—
9	Lianhua Supermarket in Hohot, March 2004	From the Ke'erqin Company: Ribeye (Rmb16), striploin Rmb16), rump (Rmb20), shoulder (Rmb10),	From the Ke'erqin Company: striploin-S (Rmb32), flank-S (Rmb25), highrib (Rmb26), bone-in shin (Rmb25)	—
10	Carrefour Supermarket in Beijing	—	From Beijing Yuxiangyuan abattoir: selected beef cubes (Rmb32), tenderloin steak (Rmb70), silverside (Rmb24), local tenderloin (Rmb28), striploin (Rmb78), beef strips (Rmb40), beef slices (Rmb40), quality beef mince (Rmb32), beef mince patties (Rmb24), shank (Rmb32), flank (Rmb26)	From Qingdao abattoir: spencer roll (Rmb150/kg), thin flank (Rmb160/kg), eye fillet (Rmb158/kg), bone-in eye fillet (Rmb70) From unspecified Beijing abattoir: tenderloin (Rmb120/kg), spencer roll (Rmb98/kg), eye fillet (Rmb138/kg)
Restaurants				
11	Generic restaurants	Purchase at wet markets for wholesale prices (typically Rmb0.5 discount for purchases over 5kg).	—	—

		Low value market segment	Mid value market segment	High value market segment
12	Mid value minority type restaurants (hot pot, fattened beef, Korean, Japanese and Brasilian BBQ)	—	Small outlets purchase ex-abattoir or beef traders for similar prices as per ex-abattoir prices below. Large chains buy at lower prices direct or from integrated plants	—
13	High value HRI trade	—	—	Buy from specialised beef traders or direct from abattoirs at prices similar to ex-abattoir prices below.

Visits to the Xinfadi market in Beijing in July 2006 revealed price information shown in Row 4 of Table 6.2.[86] The visit confirms the existence of price differentials for various cuts, though not to the extent reported on the wholesale market websites. Stallholders do not sell the number of cuts listed in Row 3 of Table 6.2, the price differentials between cuts are very small and there is very little difference in the products and prices offered by different stallholders with the exception of hot pot slices.[87] This suggests that some of the details reported in the wholesale marketing system are contrived. Similar differentials were evident in a wet market in mid sized regional market in Shandong (Row 5 of Table 6.2). Price differentials between cuts were very low in a remote city in Xinjiang (Row 6 of Table 6.2).

The existence of (albeit small) price differentials for different cuts on the mass market is a significant development in the industry. Before the

[86] Many dozens of wet markets and wholesale markets have been visited on fieldwork throughout China. Some of the details of those markets are provided in previous publications such as Waldron, Brown and Zhang (2007, Section 4.5.5).

[87] The Xinfadi market has about 50 individual stallholders which sell up to 200 kilograms of beef each per day. Each stall displays cattle and sheep carcasses and butcher to customer requirements. The carcasses were all sourced from designated slaughter points in nearby parts of Shandong and Hebei or in the suburbs of Beijing. The stalls and the carcasses displayed health and inspection certification. Hot pot requires beef with various levels and types of fat that is frozen and then thawed again for slicing by machine and then packaged. The meat is more expensive because of the transformation of the product, but also because of the specific meat required, although there are various types of hot pot meat.

year 2000, it was rare to see price differentials for different cuts at all. These differentials may increase as consumers become more discerning and affluent, and as supply chains orient to filling these preferences.

However, the significance of price differentials for different cuts in the mass market is constrained by several factors. First, as mentioned above, the small price differentials between the cuts in major city markets are even smaller or non-existent is distant regional markets. Second, many of the differentials relate to the butchering process and whether the cut is bone-in or bone-out rather than any intrinsic properties of the meat. Third, the higher value cuts (e.g. fillets) make up only a small (a few kilograms) proportion of the average carcass in China, and therefore have only a small impact on the total value of the carcass ex-abattoir, which is the primary concern of slaughter units as shown in Chapter 7.

The development of price differentials will signal and provide incentives for the production of cattle that grow larger sized premium cuts. These tend to be located on the back of the animal (loins, fillets) and to a lesser extent the rump. Because of their breeding for draught and transport work, local Chinese cattle breeds have undeveloped hindquarters but well developed forequarters. Breeding programs to develop breeds with larger hindquarters are designed to "change the shape" of Chinese cattle. However, developing the size of premium cuts also requires intensive and expensive feeding regimes that are unlikely to be offset by slim price differentials in the low value mass market.

Another major development in wet markets over the past 10 years has been the improvement of food safety and hygiene conditions. Before 2000, the stalls of beef sellers in China were almost always unhygienic with very basic infrastructure. Food safety and hygiene conditions in many mass markets are still poor—especially in rural areas and county cities—but have improved significantly in larger cities. In large wholesale markets at least, there has been investment in market infrastructure including butchering equipment, washing facilities, cold storage (for meat not sold that day)[88] and better hygiene inspection of the stalls. Most significantly, much of the beef and stallholder facilities in large and regional cities is now inspected and certified, though the rigour of inspection varies. Thus, even the mass market beef is becoming more hygienic in large cities, though much less so in rural areas.

It is also important to note that wet markets provide important outlets for mid value beef. In large and even many regional cities, it is very common to

[88] This has been legislated in some cities, where cold storage is required in markets. Meat entering Beijing by truck must be refrigerated.

see stalls in wet markets that represent, or are agents or franchises of large abattoirs, as indicated in Rows 5 and 7 of Table 6.2. These stalls sell beef at nearly double the price of stallholders selling generic beef in the same market because it is well presented (usually in glass cabinets), is sourced from a large and well known brand name, or is certified under a certification scheme (for example Green Foods). Customers of this product included three and four star HRI customers, and more discerning consumers. As is the case in Hong Kong, these types of well managed wet market outlets can be expected to continue well into the future and will not necessarily be put out of business by supermarkets. Markets will remain important in China due to cultural preferences for daily fresh product and daily shopping in local markets, and low levels of car ownership and parking bottlenecks associated with shopping in larger supermarkets.

While wet markets and their upgrading represent an important part of food retailing, they also constitute a major source of employment. Statistics on the numbers of traders and stall holders that deal in individual commodities including beef are not collected in China. However, using several assumptions it can be conservatively estimated that there are more than 200,000 beef and beef offal traders and stallholders in China.[89] Returns to beef and beef offal stallholders in wet markets have not been systematically calculated but appear to be modest at around Rmb3,600 per year (returns to labour, management and capital).[90] Returns vary enormously depending on volume, type of beef, seasonality and number of

---

[89] Every township and urban district in China (42,000) is conservatively estimated to have four households that trade beef and beef offal or that run beef and beef offal market stalls in one of the 50,000 markets in China (although these operate periodically). In addition, fieldwork suggests that there are at least 100 beef intensive counties in China that have more than 150 beef and offal traders. The Ministry of Agriculture estimates that there are 4,200 "scale" (county and district level) markets in China. Visits to these markets suggest an average of eight beef and beef offal traders. The Ministry of Agriculture count another 300 "large" wholesale markets in China. Visits to many of these markets suggest that these have an average of 20 beef and beef offal traders. Together, this totals to about 220,000 beef and beef offal traders in China. Given that China produces about 47 million beef carcasses per year, this means that each trader handles about 220 carcasses per year. This is broadly consistent with fieldwork observations, where small stallholders generally handle half a carcass per day, though with seasonal variation, but where large traders can easily wholesale three carcasses. In addition, China produces more than one million tonnes of edible beef offal per year, most of which is sold through stallholders in markets.

[90] This estimate is based on the sale of one carcass per day, with a 20% margin on purchase to sales price (after all butchering, cut differentiation and value adding) and the payment of inspection, market management, power and water costs outlined in Waldron, Brown and Zhang (2006).

days operating, and the degree of integration with slaughter operations and practices. The low returns help explain the widespread use of illegal and speculative practices by stallholders (such injecting the beef with water) and low investment in hygiene (such as stainless steel butchering equipment and washing facilities).

6.2.1 Prices in supermarkets

Supermarket sales are reported to be growing at up to 30–40% per year (Hu *et al.*, 2004),[91] with 1,083 supermarket chain stores in Beijing alone in 2004 (Beijing Statistical Yearbook, 2005). Every regional city in China now has supermarkets. However, the growth is occurring from a low base and traditional wet market channels discussed above remain entrenched and by far the most important retail channel for beef and many other fresh food products, even in major cities. Chinese consumers prefer to shop regularly (often daily) for fresh product in wet markets. As an indication of what could be expected in mainland China, supermarkets are not the dominant form of retailing in Hong Kong and Taiwan, where supermarkets have long been present and which have more affluent populations (Goldman *et al.*, 1996; Finch and Longworth, 2000). Based on extensive surveys, Goldman and Vanhonacker (2006) and Goldman (2003) discuss the major constraints to development of "Western style" supermarket sector in China and report that all have abandoned "modern supply chains" including contracts to source product, preferring instead to rely on dealers. The commodities of pork and fruit were the exceptions, while dairy products could be added to this list.

As is the case with wet markets, there is also a large continuum in the type and price of beef sold in supermarkets, which source product through a range of supply chains. As with other food products,[92] there is often little or no difference in the type of beef offered between wet markets and supermarkets. For example, in a Hualian supermarket in Ji'nan City in Shandong, a glass cabinet displayed a large rump of beef that was derived

[91] Based on a survey in China's six largest cities Goldman and Vanhonacker (2006) estimated that supermarkets sold 79% of packaged and processed foods, 45% of meats, 30–40% of poultry, fish and fruits, and 22% of vegetables.

[92] It is well known in industry and government that much of the fresh product sold in most supermarkets is sourced from wholesale markets, and that claims of direct and controlled/certified purchases by supermarkets are widely overstated. Wholesale market open earlier in the morning than supermarkets, giving supermarket purchasers time to buy fresh food early in the trading day.

from a market or from a designated slaughter point—not an abattoir—at an advertised price of Rmb22 per kilogram (see Row 8 of Table 6.2).[93]

The retail price of the beef in the supermarket was higher than at wet markets, as is necessary to offset the higher cost structures of supermarkets. Consumers are also willing to offer modest premiums for food safety and presentation in the supermarket environment. However, the upstream industry participants (slaughter units and cattle producers) almost certainly did not receive high revenues from the beef sold in the supermarket. In the Ji'nan supermarket mentioned above, next to this beef from an unmechanised slaughter point was better quality beef that sourced from a mechanised abattoir and sold for a slightly higher price.[94] Given these small premiums, and the significant extra costs in selling through a supermarket, it is difficult to see how these supermarkets could pass back significant premiums to producers.

However, some supermarkets sell beef at a price that allow premiums to be passed back to abattoir level. For example, a large supermarket in Hohhot City in Inner Mongolia that belongs to the Lianhua chain contained a section run by the Ke'erqin Company, one of the most famous beef companies in China. Specific cuts were well presented in trays at what could be regarded as "mid value" prices (see Row 9 of Table 6.2).

A small number of supermarkets sell high value beef to wealthy Chinese and expatriates. For example, a Carrefour supermarket in Beijing had an area with well presented refrigerated display cabinets that contained attractive looking beef packaged in trays and cellophane in small portions for a single meal or in two steaks weighing a total of between 200 and 300 kilograms. Above the cabinets were signs advertising beef. The advertising information included the regions of origin (Beijing or Qingdao), while the highest priced product was marketed as coming from a well known abattoir in Beijing (Yuxiangyuan). The advertising mentioned general attributes of the beef (flavour, juiciness and freshness) and was sometimes misleading (that the

---

93 The muscle was dark and the fat was yellow in colour. The dirt on the carcass showed that the animal was slaughtered by hand on the ground (bed slaughtering) at an unmechanised designated slaughter point. There was health and inspection certification attached to the beef and specifying the source of the beef. While customers are unlikely to be able to recognise the name of the slaughter place, it was known to industry experts present on fieldwork that there were no mechanised abattoirs in the area.

94 The beef was labelled by a large well known company called Hengdian. The beef was presented in 25 kilogram chunks and cut to customer requirements. The fat was off white and the beast was probably finished on grain and slaughtered at around two years of age.

cattle were Angus breed and raised on special diets). The price schedule for this supermarket is shown in Row 10 of Table 6.2. Of most interest is the large range of products available and the large price differentials for the products.

6.2.2 Prices in restaurants

As is the case for markets and supermarkets, the Hotel, Restaurant and Institution (HRI) trade is very diverse and covers the full continuum of low, mid and high value beef. It is not possible to examine the enormous range of HRI outlets in any details but some relevant issues are as follows.

The majority of beef consumed in the HRI trade is purchased by small street side restaurants as low value, generic beef from wet markets (as indicated in Row 11 of Table 6.2). However, this generic beef is suitable for many of the heavily sauced, spiced, stewed or boiled dishes on Chinese dishes, which require few specific quality attributes. It is also suitable for many of the minority restaurants (Xinjiang skewers, Hui Muslim noodles, and some Korean dishes) and fast food chains (e.g. McDonalds). Although these restaurants add to overall beef consumption, they provide little opportunity to increase margins for producers. In some cases, however, the restaurants will buy particular cuts of beef (e.g. fillet) for particular dishes (e.g. sizzling iron beef).

The mid value out-of-home HRI trade provides genuine scope for higher margins as suggested in Row 12 of Table 6.2. They lie somewhere between a cheap fast food outlet and an expensive specialist restaurant. The HRI outlets where beef is a major part of the menu tend to be "ethnic" type of outlets, including mid value Korean and Japanese barbeque restaurants, Brasilian barbeque and American fattened beef restaurants. Virtually every large and regional city in China has at least one and often many more of these types of restaurants. However, the most import type of restaurant for beef (and even more so for sheep meat) are hot pot restaurants. There are many thousands of hot pot restaurants in China, many of which are franchises in large chains. Some of these restaurants sell 100 kilograms of beef per day. There is often five different types of beef dishes on the menu in hot pot restaurants so the requirements very considerably. However, the beef usually has some marbling and the cattle are finished on grain. The beef is purchased from processors in rolled, frozen form. It is served sliced usually in 350 gram portions that sell for between Rmb20 and 40 each portion. However, these prices reflect restaurant costs, including up to

70 staff per restaurant. For more details on the hot pot trade see Waldron, Brown, Longworth *et al.* (2007, Section 7.3).

This mid value beef is supplied through various supply chains. As mentioned above, smaller HRI outlets can buy safety and quality assured product in sufficient quantities through "upmarket" stalls in wet markets. However, in restaurants where beef constitutes an important part of the menu, there are that often more direct sources of supply. For example, a mid range Korean restaurant will have direct relationships and ongoing orders with a particular slaughter unit. These slaughter units are more likely to be small (or old) local abattoirs or certified centralised slaughter points that can supply particular cuts, than they are to be large modern abattoirs. In the cases of large franchise chains (especially hot pot), product is usually sourced centrally by the brand owner. In the case of Xiaofeiyang and Caoyuan Xingfa, product is sourced predominantly from old General Food Company abattoirs. In the case of more upmarket hot pot restaurants (such as Yuxiangyuan), the beef is sourced through vertically integrated modern abattoirs.

At the high end of the HRI market are the exclusive outlets (see Row 13 of Table 6.2) that tend to cater for expatriate Japanese, Koreans or Western customers. There are too many specific requirements and product attributes in the high value markets to list here but some generalisations can be made. First, most of the beef demanded has to be tender, contain some extra-muscular fat, sometimes marbling and the cuts required (especially loins and fillets) must be relatively large in size. Production of this type of beef can be technically demanding and high cost at both slaughter and cattle production stages. In addition to breed selection, cattle must be fed a significant period of time in feedlots and these costs must be offset by price premiums. Steaks that weigh about 200 grams sell in an exclusive hotel restaurant in Beijing for Rmb150. However, these cuts form a small portion of the overall carcass and prices include large margins from the restaurants. Due to a lack of quality, consistency and adherence to grading schemes in China, a large proportion of the beef for premium markets is imported (Waldron, Brown and Zhang, 2007).

6.3 Prices ex-abattoir

Retail prices discussed above provide a guide to the way that the final beef market is segmenting in various retail outlets. However, the retail prices reflect not only the prices and attributes of the beef bought by the retailers, but also the costs of the services that the outlets provide including butchering, preparation, service and convenience as well as the profit margins for these service providers. In order to gain a better idea of beef prices that exclude these retail marketing costs and margins—that is, the derived demand for beef at the slaughter level—prices for beef at point of sale out of the various slaughter units are listed in Rows 1 to 3 of Table 6.3. The prices in Table 6.3 are offer prices and will vary somewhat from the actual prices, depending on the size of the order, payment terms, trust and personal relationships.

In addition to showing the large range of products and prices produced by slaughter units, the price schedule for the abattoir prices in Row 3 of Table 6.3 also provides an insight into the difficulties of synthesising discrete value chains. This is further illustrated in the oblique dotted arrows between slaughter units and retail outlets in Figure 3.1 where any one modern abattoir supplies different cuts into low, mid and high value end markets. Similarly, the best cuts from even the most rudimentary slaughter household can be sold into mid value markets. It is therefore difficult to trace product flow through distinct (low, mid and high) value chains. In addition, because the composition of different cuts and the quality of the beef produced varies enormously by animal, it is also very problematic to use the price of the various cuts to derive the gross value of a carcass.[95] As a result, several abattoirs and industry consultants involved in beef projects in which the author was involved base slaughter output values on an average price for the beef for an entire carcass. An accepted value of low value beef (bone-out) in 2006 was Rmb15/kg, mid beef was Rmb21/kg, and high beef was Rmb27/kg, as shown in Row 4 of Table 6.3. In Chapter 7, these values are used as output values in the budgeting of various slaughter units listed in Row 5 of Table 6.3.

[95] As a general rule, of the meat on a lot fed animal with a carcass weight of 220 kilograms, less than 3% is very high value, premium cuts (striploin and thin flanks). Another group of prime cuts (eye fillet and tenderloin) account for less than 7% of the total meat and commonly sells at a slightly lower price of around Rmb70 per kilogram. Other cuts for Western dishes and chuck meat for Chinese dishes can be worth between Rmb40 and Rmb50 if good quality. Brisket, rump and other major cuts can be worth Rmb30 if good quality and can weigh up to 70 kilograms. Other cuts that comprise the majority of carcass are sold on generic markets and as off-cuts.

Table 6.3: Prices at slaughter unit level for low, mid and high value market segments

		Low value market segment	Mid value market segment	High value market segment
1	Slaughter household in Shandong servicing Beijing market	Sell to trader for Rmb14/kg in half carcass form bone-in		
2	Designated slaughter point	Rmb14.8/kg in half carcass form for unmechanised slaughter point with cold storage and delivery services	Rmb20–30 for particular cuts	
3	Shandong modern abattoir	Chinese cuts: Standard fattened beef (Rmb13), No. 3 fattened beef (Rmb13), Chuck tender with membrane (Rmb18), Outside (Rmb19), Silverside (Rmb19), Knuckle (Rmb19) chuck tender no membrane (Rmb20), Eye round (Rmb22), Shank (Rmb22), No.2 fattened beef (Rmb24)	Western, Korean and Japanese cuts: Chuck steak (Rmb25), Tri-tip/ bottom sirloin triangle (Rmb25), Brisket (28), Bone-in brisket (B) (Rmb28), Blade (Rmb30), Brisket point (Rmb32), Bone-in brisket (A) (Rmb34), T bone (Rmb45), Short ribs (Rmb50) Chinese cuts: Fattened beef brisket (Rmb32), F striploin (Rmb34), No.1 fattened beef (Rmb34), American fattened beef (equivalent to Grade 1 Chinese fattened beef) (Rmb35), B chuck roll (Rmb36), B striploin (Rmb36), B eye fillet (Rmb45), A chuck roll (Rmb48), A striploin (Rmb48)	Western, Korean and Japanese cuts: B eye fillet (US cut) (Rmb65), S tenderloin (Rmb65), A tenderloin (Rmb65), B tenderloin (Rmb65), A eye fillet (US cut) (Rmb70), Chuck eye log (Rmb70), S eye fillet (US cut) (Rmb75), S special striploin (Rmb90), Sorabol (combines boneless Spencer roll with the chuck roll to form a large cut (>11kg to reach grade A) used for Korean BBQ) (Rmb110), S striploin (Rmb110), Thin flank (Rmb190) Chinese cuts: A eye fillet (Rmb60), S chuck roll (Rmb65), S eye fillet (Rmb80)
4	Conversion to ex-abattoir price over whole carcass bone-in (parameters used for slaughter unit outputs)	Rmb15	Rmb21	Rmb27
5	Serviced by	• Slaughter households • Designated slaughter points • Low value cuts from abattoirs	• Particular cuts from better equipped designated slaughter points • Mid value cuts from abattoirs	• Premium cuts from mechanised abattoirs

Chapter Seven

Development of the processing sector

Having analysed low, mid and high value market segments and retail outlets in the previous chapter, this chapter examines the suppliers which, in the case of the meat sector and beef industry, are mainly slaughter units.[96] Different types of slaughter units tend to operate in different value agrifood chains but there is also some cross over. In order of importance,[97] the slaughter units include: uncertified individual slaughter households (where product is sold into low value markets); certified centralised slaughter points (which can sell some product into mid value markets); and modern mechanised abattoirs (which can sell some product into high value markets). These different forms of slaughter units are significant employers, employing at least 150,000 people, mainly in rural or suburban areas.[98]

As discussed throughout the chapter, while uncertified household slaughtering is formally banned in China for food safety reasons, there has

---

[96] There are also large numbers of beef and beef offal traders and processors engaged in the further transformation of beef (into meals, soups and cooked beef product) that buy from slaughter units and sell to retailers. Import agents can also distribute independently from the slaughter units.

[97] Data on the numbers and proportions of each type of slaughter unit in China are not systematically collected, although various estimates have been made. In the late 1990s, it was estimated that in Henan 4.2% of cattle were slaughtered in mechanised abattoirs, 13.5% were slaughtered in specialised villages, and the remainder (about 82%) were slaughtered by individual households outside the specialised villages (The Study Group of the Chinese Association for the Promotion of International Agricultural Co-operation, 1997). As discussed below, there has been some increase in the former two categories and a decrease in the latter category, but slaughter households probably still kill around 70% of the cattle in China, and of the remainder, 20% are killed in designated slaughter points at county level and below and 10% in abattoirs at prefecture and province level. Even Ji'nan City—the capital of Shandong with over five million people—received about 70% of its beef from slaughter households in nearby Qihe County in 2006.

[98] The number of people employed in the slaughter sector are based on estimates that there are around: 40,000 cattle slaughter households that employ at least one household member, an Ahong (Iman) slaughterer and another general worker; about 2,000 designated cattle slaughter points that employ an average of 10 staff each; and 150 city-province level cattle abattoirs that employ 100 people each. It is important to note, however, that much of the employment is temporary or seasonal in the slaughter sector.

been only localised implementation of these bans. Increasing implementation of the ban on uncertified slaughter households over the medium term future will increase the relative importance of two other types of slaughter units, namely certified centralised slaughter points and modern mechanised abattoirs. Policy makers see the modernisation process as a movement up this continuum of structures. However, this desired process is ultimately determined by the relative profitability of the various types of slaughter units. This relative profitability is a function of the cost structures of the slaughter units, especially the input costs of their slaughter cattle, the efficiency by which cattle are transformed into different types of beef, and the price at which the different beef outputs can be sold. Thus, developing budgets for the different types of slaughter units and testing the sensitivities of these budgets to cattle input and beef output prices and product attributes provide important insights into future developments in the slaughter sector. Analysis suggests that policy to expedite the development of large modern mechanised abattoirs is not aligned with market realities and is not therefore sustainable.

The structure of the chapter is illustrated in Figure 7.1, which is an extract from the main study framework (Figure 3.1). The analysis draws on partial budget analysis of various representative slaughter units, through the process explained in Box 1.

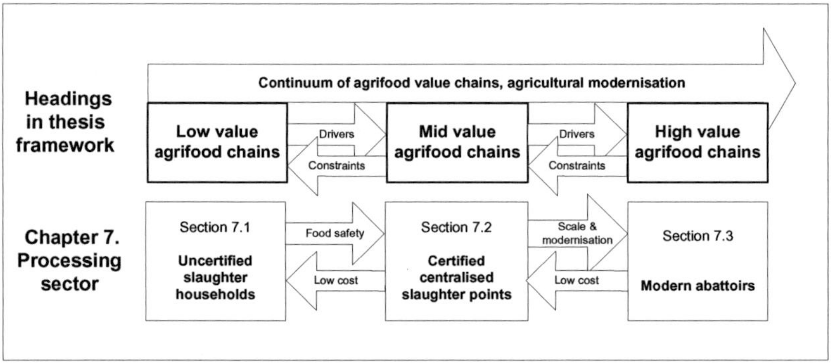

Figure 7.1 Outline of structures and drivers and constraints to modernisation of the slaughter sector

7.1 Uncertified slaughter households

The majority of China's cattle slaughter and beef production takes place within individual households that are not certified, registered or regularly inspected. These processing structures are inconsistent with central level food safety or environmental objectives in China. The budget analysis presented below suggests that the costs of meeting food safety and environmental standards would be significant for individual slaughter households. Even if theses costs could be met by the slaughter household, the transaction costs for inspection agencies of inspecting large numbers of individual slaughter households would be very high. Furthermore, local inspection agencies have few avenues for recouping these costs, given measures over the 2000s to abolish taxes and cut inspection fees for many activities related to agriculture and food safety, including slaughter. The banning of uncertified household slaughtering raises a series of social issues for policy makers, especially with regard to the livelihood impacts on Hui minority people that dominate cattle (and sheep and goat) slaughter. More information on these and other issues in the cattle slaughter sector is presented in Brown *et al.* (2002a).

Box 1. Background to the abattoir budget analysis

The partial budget analysis of abattoirs in this chapter is based on a more detailed analysis reported in Longworth *et al.* (2001, Chapter 9). Formal surveys of abattoirs were not conducted due to commercial and policy sensitivities in the sector. However, more that 40 abattoirs were visited in the late 1990s as part of the original study to obtain detailed information on costs, revenues, margins and incentives facing slaughter operations. Semi-structured interviews were conducted with abattoir managers and information cross-referenced with information derived from officials, industry consultants and other industry participants from cattle supply, slaughtering and processing, and beef distribution sectors. Another 15 cattle and sheep abattoirs have been visited since the year 2000.

The analysis in this chapter used the budgeting process reported in Longworth *et al.* (2001, Chapter 9) and retained many of the original technical parameters. However, much of the data has been updated in this analysis, including prices reported in the cattle input-beef output sensitivity tables. Relationships between cattle weight, dressing percentage and per unit price premiums or discounts are based on values reported by abattoirs on fieldwork and technical reports including Zhou *et al.* (2001) and Jiang (1995).

The budgets are also used to test the sensitivity of abattoir returns to variations in the major inputs (slaughter cattle) and outputs (beef) of the abattoirs as shown in Table 7.3 and Table 7.6. Beef output price parameters were established as an average value for the entire carcass rather than an aggregation of many different value cuts for reasons established in Section 6.3. Parameters for cattle inputs include prices and attributes of the cattle including live weight and dressing percentage. These parameters are consistent to those used in Chapter 9 (Table 9.3 and Table 9.5) which include other information on cattle inputs including age of animal and time on feed. The relationships between the slaughter cattle inputs of the abattoirs and the slaughter cattle outputs of the cattle fattening units can be used to synthesise a major part of the beef cattle supply chain.

Input-output sensitivity analysis was selected as the only form of sensitivity analysis to be reported on in any detail for several reasons. First, the relationship between inputs and outputs is the largest determinant of abattoir profitability. Cattle costs are by far the largest cost item for the abattoirs and beef outputs constitute by far the biggest revenue item. Second, it has not been possible in the confines of this single chapter to report on other factors that impact on abattoir profitability—such as capacity utilisation, labour costs, capital costs, and butchering processes—although these are overviewed in the text and in more detail in Longworth *et al.* (2001, Chapter 9). Third, the focus on inputs and outputs highlight major issues under review in the broader study, namely value chain segmentation and integration.

It is argued that while the budgeting and sensitivity analyses provides useful insights into issues in the abattoir sector and its role in the beef supply chain, the analyses produce representative and indicative results only. Thus, the results have been interpreted cautiously and only to provide additional supporting information to the main discussion in the chapter.

7.1.1 Issues in the household slaughter sector

The dominance of individual household processors in the beef (and sheep and goat meat) industry differs to the processing sectors of most other agricultural commodity chains, which tend to happen on a larger scale in more centralised facilities. For example, pig and chicken slaughter, grain and feed processing, wool, cashmere and cotton processing and fruit and vegetable processing is usually done by companies or collectives on a scale above the individual household level. While individuals and groups of households dominate fruit and vegetable assembly, sorting, washing, packing, but there is little "processing" involved.

Individual household processors and traders in rural areas play a valuable role in agricultural and rural development. They source agricultural inputs and process them at low cost in a form and price that meets the demands of mass markets, especially given the low willingness and ability to pay in those markets. In the case of beef and other agricultural industries, the supply chains can be short and direct with close integration (through family and kinship members) from production to retail levels, but can also consist of a long series of household participants. Small scale processing also provides a major source of employment at a local level. It provides opportunities for households to "step out" of marginal on-farm agricultural activities to more specialised and entrepreneurial off-farm processing and service activities. There may be up to 40,000 cattle slaughter households in China,[99] which are distributed throughout rural areas, but can also be highly concentrated in certain pockets.[100]

Policy makers are, however, concerned about the negative aspects of small scale household processing. From a fiscal perspective, individual slaughter units contribute little to local government revenues because they are too small and dispersed to monitor and because slaughter taxes were abolished from 2002. In this regard, government would prefer to foster enterprises that can be taxed, with a proportion of tax revenues retained locally. Environmental problems include the runoff of untreated effluent (blood, entrails) into public areas and water courses (Yan, 1999).[101] Of most concern in the slaughter sector are food safety issues, which have received an enormous amount of attention in China in the wake of the outbreaks of Severe Acute Respiratory Syndrome, Avian Influenza, melamine contamination in dairy, various pig diseases, as well as Bovine Spongiform Encephalopathy overseas. Meat can also contain traces of disease and residues from live animals (such as Bovine Spongiform Encephalopathy, Foot and Mouth Disease and tuberculosis) and production practices (growth hormones, antibiotics, drenches, external parasite treatment).

The slaughter sector is a critical juncture in food safety aspects of the livestock supply chains for several reasons. The inspection of live animals (for disease), carcasses and organs (for disease and residues)

---

99 This estimate is based on the assumption that every rural township in China (not street committees) have an average of just one cattle slaughter household (35,000 households), and 50 intensive counties have 50 slaughter households. This estimate is broadly consistent with the estimate that households slaughter an average of 70% of Chinese cattle, at an average of 880 head per year, a throughput which aligns with fieldwork observations and interviews.

100 For example, in Yangxin County in Shandong, about 1,000 specialised households are located in 16 specialised slaughter villages. Meitang Town in Linyi County in Shandong has five Hui ethnic minority villages where almost every household in the village is a specialised slaughter household. Dachang and Sanhe in Hebei counties on the border of Beijing are also well known examples.

101 There are parallels in other industries. For example, small scale leather tanneries, paper mills, feed mills and even coal mines were forced to amalgamate or close in 2001 under an edict of the State Economic and Trade Commission, as a means of building critical scale and waste management facilities and system.

occurs at slaughter stage. There is also large potential for unhygienic practices at the slaughter stage (bacterial and microbial infection, sanitary and phytosanitory problems). Small uncertified slaughter households use mainly bed slaughtering techniques (rather than on hooks) on dirt (rather than concrete) floors and have highly variable procedures for washing and sterilisation of slaughter areas, equipment and meat. There are rarely cold storage facilities and distribution methods are crude.

Enforcing food safety and environmental regulations in a processing sector dominated by small, fragmented, individual households is highly problematic. It is not logistically possible for authorities[102] to regularly inspect individual slaughter households and, in many cases, State (as opposed to collective) authorities do not even know where and how many slaughter household operate. Even if it was possible, inspection of the dispersed households would impose very high inspection costs on State inspection agencies that operate under tight budgetary constraints. Township, village and peer group pressure is not applied systematically enough to maintain food safety standards on an informal basis.

As a result, household livestock slaughtering is legal only in certified centralised slaughter points as discussed in the next section. The policy has been applied to pigs and chickens but only in some areas for cattle and sheep for several reasons. First, as the major type of meat classified as a staple food across China, the food safety of pork is given higher priority than is the case of beef. Second, the slaughter of chickens generally takes place in a more centralised manner than is the case of cattle. Having said this, however, a lot of pigs and chickens are still slaughtered by rural households and in unhygienic conditions.[103]

Third, socio-political factors also explain why authorities have traditionally been reluctant to enforce hygiene, inspection, registration and certification demands on cattle, sheep and goat slaughter households. As shown in Table 7.1 the low margins in household slaughtering means that the extra costs of inspection and infrastructure upgrading would, if imposed, impact significantly on the slaughter households. The impacts on cattle and sheep slaughter are felt most strongly by Hui minority people that are said by the China Cattle Association to still constitute up to 90% of the

[102] This includes the Animal Husbandry Bureau (livestock disease inspection station), the Bureau of Internal Trade, the Ministry of Health and the Industry and Commerce Administration Bureau.

[103] Even in the more tightly regulated pig slaughter sector, it is known that many uncertified pig slaughter households remain active by marketing their pork through certified plants.

cattle, sheep and goat meat slaughter, trade and retail sectors (Shen *et al.*, 2008). Han Chinese authorities throughout China have traditionally been reluctant to take actions that are perceived to harm local Muslim, especially Hui minorities.[104] However, this problem is not insurmountable and can be achieved through a medium term adjustment policy. In particular, the centralised certified slaughter point model described below in Section 7.2 provides a mechanism for the absorption of Hui people into safer slaughter systems, if managed appropriately.

7.1.2 Budget analysis of representative slaughter household

In order to gain more insight into issues in household slaughtering and policy, this section presents a partial budget of a representative slaughter household in the Central Plains of China, which is summarised in Table 7.1. The supply chain by which the representative slaughter household buys inputs and sells outputs is depicted in Figure 7.2, which is extracted from the main study framework. The data is sourced from many visits to individual slaughter households and corroborated with various authorities and government officials familiar with the sector.

The representative slaughter household kills 2,000 head pear year, or an average of 5.5 head per day, though with daily and seasonal variation.[105] The household purchases a 312kg animal (from the specialised fattening households in Chapter 9) at Rmb8 per kilogram liveweight with a dressing percentage of 50%. Carcasses are sold at an average price of Rmb13/kg in half carcass form. The half carcass is sold to beef dealers or stallholders who then butcher the carcass and sell partially boned out beef at prices and differentials quoted in Chapter 6. The purchase costs of cattle are actually higher than the sales revenues from the beef carcasses alone but the shortfall is made up through the sales of hides (Rmb390 per beast) and offal (Rmb50), both of which are sold to specialised dealers.

[104] This issue has been cited numerous times on fieldwork and in documents such as that by the Shandong Industry and Commerce Administration Bureau (2006). Hui people have historically had intermittent areas of conflict with the Han, and are sometimes described as stubborn and vocal. While they constitute only 0.6% of the population in a province like Shandong, 1.28% of the population in Dezhou Prefecture, and 10% of the population of Yangxin County, they are the only recognised minority people in these areas.

[105] Households sometimes slaughter every second day, with higher slaughter numbers in winter and lower in summer.

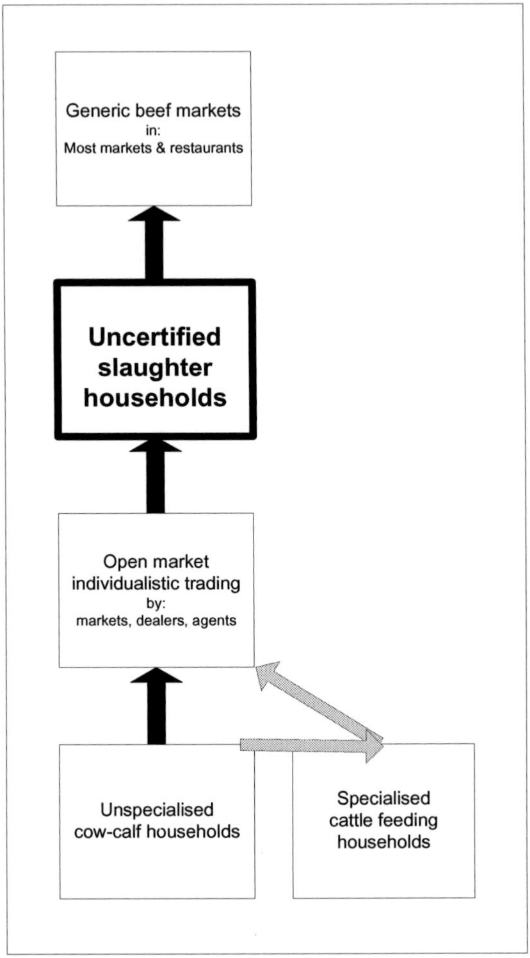

Figure 7.2: Synthesised supply chain for the representative uncertified slaughter household

Table 7.1: Partial budget of slaughter household per head costs and revenues in Central Plains, 2006

	Rmb/head	%
Revenues	**2,593**	**100%**
Beef sales (CW bone-in)	2,028	78
Hide	390	15
Offal	175	7
Costs	**2,524**	**100%**
Cattle purchase (LW)	2,496	99
Cattle transport	13	1
Employed labour	12	0
Depreciation on infrastructure	1	0
Overheads	3	0
Returns to household labour, management and capital (without registration and inspection costs)	**69**	
Registration and inspection costs—if applied	**29**	**100%**
Registration	12	41
Facilities hygiene inspection	4	14
Cattle disease inspection	5	17
Carcass inspection	5	17
Capital investment to meet hygiene standards	3	10
Returns to household labour, management and capital (with registration and inspection costs)	**40**	

Slaughter costs for these slaughter households are low. Two types of labour are employed from outside the slaughter household. A general worker is employed full time at a cost of Rmb7,000 per year. In addition, many slaughter households employ on a casual basis an "Ahong" slaughterman for Halal slaughtering, who kills and skins the animal in accordance to custom at a flat rate of Rmb8 per beast.[106] These slaughtermen often works in teams of two or three which are also employed on busy slaughter days, and rotate between various slaughter households. In addition to the employed labour, household labour includes the household head (for sourcing cattle, beef marketing and management) and all other members of the family (for tasks like sorting and washing). Costs for depreciation[107]

[106] The rates paid to Ahong slaughtermen are high because very few are registered in a given county or city. Ahong slaughtermen are must undergo religious and technical training and are registered by United Front Work Department (*Tongzhan bu*) of the Chinese Communist Party.

[107] Depreciation costs are applied to the infrastructure and equipment items of a slaughter shed, holding yards, rails and hooks, well, pump and slaughter equipment.

and overheads[108] are low especially relative to other types of slaughter units and provide their source of competitive advantage in supplying low value markets.

Under the set of assumptions and parameters outlined above, returns (to household labour, management and capital) are modest at around Rmb69 per beast. However, when multiplied over a slaughter of 2,000 head per year, household returns are Rmb137,000, which is far higher than the representative unspecialised cattle producing household and higher than the representative specialised cattle fattening household (in Chapter 9). In addition, the returns per beast are higher than those of the representative modern abattoir, which are negative. Returns can be higher for slaughter households that are integrated into the beef retail stage, which internalise butchering and other margins and costs within the households/group structure.

These estimates do not, however, include a range of health and inspection fees and the capital investment costs required to meet the standards.[109] As discussed above, these fees and costs are not applied to uncertified slaughter households in most of China. If they were applied as in the case in some parts of China, returns would be reduced by about 40%. Officials argue that the investment in infrastructure required to meet the national standards would be particularly onerous for slaughter households.[110] Nevertheless, the budget analysis suggests that these costs would be able to be absorbed by some slaughter households.

7.2 Certified centralised slaughter units

The next step up in the continuum of slaughter units are certified slaughter units, which are more commonly known in China as "designated slaughter points" (*dingdian tuzaichang*). This category covers a wide range of

---

108 Includes phone and power costs.

109 After 2002, slaughter taxes (which were rarely levied on individual slaughter households anyway) were abolished.

110 In the late 1990s, the Bureau of Internal Trade regulated that slaughter households have a registered capital investment of Rmb30,000 but this regulation was never enforced. It is estimated that today for the representative household, a capital investment of Rmb100,000 would be required to meet standards to sell into Beijing. Households investing in cold storage and transport infrastructure killing 20 head per day would require about Rmb500,000, although this would allow access to mid value and more distant markets, according to interviews with some slaughter units.

slaughter units but they also share common characteristics. They generally produce safer meat than uncertified slaughter households, have lower cost structures than large modern abattoirs, can absorb large numbers of rural slaughter workers, and have the potential to sell at least some and sometimes much of their beef in to mid value markets. As a result, slaughter units that fall into this category can be expected to play an increasingly important role in the development of the beef industry.

7.2.1 Issues in certified centralised slaughtering

As mentioned above, the slaughter of pigs in household courtyards and on roadsides is formally banned and the ban has been implemented on a widespread scale which has moved slaughter into more centralised certified operations that are notionally subject to inspection.[111] These areas are known in China as designated slaughter points of which there are many tens of thousands across China.[112] It is only a matter of time—probably bought about by a major cattle disease or beef contamination outbreak which has not been detected at all or in a timely manner because of the difficulties of inspecting "the multitude" of "fragmented" slaughter households—before a ban on uncertified household slaughtering is formally extended to cattle and sheep across China.

---

111 The major measures to improve food safety aspects at the slaughter level are the "PRC Animal Disease Prevention Law", the "Pig Slaughter Management Regulations" (promulgated in by the Ministry of Internal Trade in 1997 and revised at local levels in 1997 and 1998) and the pig slaughter and butchering workshop design standards (GB50317-2000). A series of steps, inspections and documents are required to register the designated slaughter points and the workers with the county Industry and Commerce Administration Bureau and the county Economic and Commercial Management and Planning branch. Registration needs to be renewed each year.

112 For all livestock types, there are about 50,000 designated slaughter points registered in China, of which 40,000 are county level and below and 10,000 are at the city or prefecture level (which are mainly mechanised abattoirs). Beijing had 67 designated slaughter points in 2004, Shanghai had 52 and Xi'an had 102 (China Meat Association, personal communication, 2004). There and there are parallels of relatively centralised processing activities in other livestock products. For example, milking and milk processing has become relatively centralised, partly because of food safety issues (bacteria in milk, hygiene of processing plants) and partly because of the need to assemble a critical mass of product.

In the meantime, the decision about whether to extend the bans to cattle and sheep slaughter is taken at local levels.[113] Authorities in many of the more developed cities have implemented the bans, partly because of higher food safety expectations (and higher willingness to pay) in these areas and because displaced labour is more readily absorbed into other sectors. Furthermore, the ban has been thoroughly implemented in Xinjiang and Inner Mongolia, where authorities have a particular responsibility to ensure the safety of beef (and sheep and goat meat) because they are classified as staple foods for minority (Uyghur, Mongolian, Hui and Kazakh) populations. In these areas, food safety concerns are prioritised over other concerns of, for example, traditional slaughter communities servicing urban areas. However, the ban does not apply to the self slaughter of livestock for own household consumption in pastoral areas.

There are perhaps 2,000 certified, centralised (or "designated") cattle slaughter units in China ranked at county level and below (China Meat Association, personal communication, 2004). Within this model, however, there are many different forms of certified centralised slaughter units.[114] Two categories of certified, centralised slaughter units are analysed in this section. The first category—called unmechanised centralised certified

[113] Uncertified individual household cattle slaughter has been banned in whole municipal provinces (Beijing, Shanghai), cities (Hangzhou, Dalian, Hohhot, Urumqi, Zhengzhou), prefectures (such as Altai in Xinjiang and Qinyang in Gansu) and counties (Ji County in Tianjin, Gong County in Sichuan). A new abattoir in Qihe County in Shandong was being built in 2006 to replace household slaughtering for Ji'nan City which supplied 60–70% of the beef sold into the provincial capital city. In provinces like Gansu that have "grabbed" the slaughter regulations for implementation, even remote and poor areas like Qingyang Prefecture had notionally implemented a slaughter ban for all livestock types by 2008.

[114] For an understanding of how diverse designated slaughtering is, consider the following examples. A designated slaughter point visited in Henan was run by the village collective on a very small scale with simple infrastructure. Authorities in Henan are trying to concentrate the slaughter of cattle within a township into a singe designated slaughter point. A large designated slaughter point in Daxing District (one of seven designated cattle slaughter points in Beijing) is collectively run by a Hui community with basic infrastructure (pens, slaughter areas) that covers a very large area. It was to be absorbed into a modern abattoir (Kangle) but which failed to proceed. In the case of Xinjiang and Inner Mongolia, most of the designated slaughter points are based in old sometimes refurbished General Food Company abattoirs. In Urumqi an unmechanised designated slaughter point owned by an enormous private company called Hualing slaughters 4,000 sheep slaughtered per day and 700 cattle to provide about two thirds of the city's beef and sheep meat, which is sold through markets also owned by the company. The company is paid a service slaughter fee and takes ownership of the skin/hides from the slaughter. A similar model is used in Changchun City, run by the Haoyue Company, which is attempting to control city meat supplies from its mechanised plants.

slaughter points—are often collectively owned, operate in a common area, with common infrastructure and common inspection regimes, that still supply predominantly to mass markets but with the potential to supply some product into mid value markets. The second category—called mechanised certified abattoirs—are larger corporate abattoirs that have mechanised slaughter lines and significant cold storage, washing and water treatment facilities. This latter form of abattoir has the potential to access mid value markets for a significant proportion of their product.

7.2.2 Unmechanised centralised certified slaughter points

The first category of centralised certified slaughter unit—unmechanised centralised certified slaughter points—kill between 5,000 and 15,000 head of cattle per year. They have a number of common features. First, they tend to have simple infrastructure such as hooks, butchering areas, livestock pens, washing facilities and sometimes cold storage facilities. Second, these facilities are often collectively owned and used by a substantial number of households and employees[115] although there are examples where the certified slaughter points were privately owned. Third, certified centralised slaughter points are registered and have at least some form of inspection system in place.[116] Fourth, the cost structures at the slaughter points are very low. While detailed data was not collected, the low cost structures are reflected in the very low service slaughtering fees charged.[117] Fifth, most slaughter points are only really able to access low value mass markets for the majority of their beef due to infrastructure, technical and skill constraints. There are, however, examples of small entrepreneurial slaughter businesses that are located on

[115] It is common to see about 10 people working in the larger township level designated slaughter points.

[116] Livestock disease inspection station officers notionally inspect at three points in the slaughter process: (a) they check disease certificates on arrival of the live cattle and inspect the animals; (b) they inspect carcasses during slaughter (meat, organs, offals); and (c) they inspect and stamp the carcasses with an edible dye. Another certificate is issued which clears the meat to be taken to retail markets and in some cases carry an identification mark and number to increase traceability. Inspection fees in Altai City in Xinjiang were Rmb2 for sheep and Rmb5 for cattle, but can be higher than this in larger areas (up to Rmb20 per head).

[117] Service slaughter fees (paid in cases where the cattle supplier retains ownership of the beef) were on all occasions lower than Rmb20 (not including inspection). However, in some cases (especially Western China), these slaughter fees are reduced by the provision of subsidies from local (especially city) government as a way of reducing or controlling prices for minority populations.

the outskirts of major cities (e.g. Beijing and Qingdao) that access mid value and niche beef markets.[118] Sixth, the centralisation of the slaughter activities allows for more centralised and larger scale assembly and marketing of whole carcasses, particular cuts, hides and offal. Thus, average per carcass revenues are likely to be higher than is the case for individual slaughter households, while transaction and marketing costs can be expected to be lower.

The points above suggest significant benefits in the centralised certified slaughter point model in meeting China's food safety requirements at a relatively low cost while also enabling continued employment in the slaughter sector by Hui minority communities. Centralised certified slaughter points will over time become the predominant slaughter model for beef and sheep and goat meat for the mass markets. However, this model has also encountered problems in implementation that need to be addressed to play a positive role in industry, regional and rural development.

First, even though the centralised certified slaughter points at county level and below are subject to some inspection, the food safety standards are widely thought—and have been observed—to be low. Some industry participants believe that there is little difference in the hygiene standards of slaughter households and many designated slaughter points. In practice, "inspection" often involves little more than handing over the money to get the stamp/paperwork, as is widely acknowledged in the industry and by increasingly cynical consumers. Low levels of food safety may actually become more pronounced if the calls of the National Congress in March 2007 to abolish inspection fees in the slaughter sector are implemented and if central government does not adequately compensate cash strapped local governments for forgoing the loss of revenue from the inspection fees. In this case, local government authorities are even less likely to implement slaughter regulations and conduct effective inspection.

The second major issue that also relates to large modern abattoirs below is that the establishment of centralised certified slaughter points has in some areas established local monopolies. This is the case in many county, prefectural and even provincial level cities (e.g. Urumqi, Changchun). In cases where there are private or corporate owners of the centralised certified slaughter points, the control sometimes also extends to the management rights of the food markets where the beef is sold. The enterprises thereby extract fees from both slaughter and from beef traders that use their market facilities (including stalls and cold storage). These companies are also known to receive preferential treatment

[118] One businessman with formal training in meat and livestock science hires an Ahong slaughterman to slaughter around 15 head per day. He ages carcasses for three to four days in cold storage and distributes the beef into Beijing in his own cold storage trucks.

on the basis that they provide safe food for minority populations. Preferential treatment includes the monopoly or oligopoly status (e.g. in a city or county) and subsidies to stabilise prices if prices exceed upper limits (set and monitored by Price Bureau) or impact strongly on consumers cost of living.[119]

7.2.3 Small mechanised certified abattoirs

With some notable exceptions, it is unlikely that the majority of beef from the county and township centralised certified slaughter points discussed above can meet the hygiene and quality requirements of mid value markets (to achieve Rmb21 average for the entire carcass in 2006 prices). However, there are many examples of small mechanised certified abattoirs with basic infrastructure that can sell at least some of their cuts on to mid value markets. For example, there are many such small abattoirs in the poor area of Zhoukou Prefecture in Henan Province that perform service slaughtering for larger meat trading companies in large cities of China.[120] Another small

---

119 With regard to designated slaughter sites Ke (2001) writes that "Because most of those sites are built on the bases of the former slaughterhouses of the state meat market companies, they sometimes use this advantage to manipulate the market. The problem is, the marketing administration authorities (the State Food Bureau system) represent the interests of the state marketing companies, the malpractice in some localities is often protected by the local authorities for market administration." While the General Food companies have undergone extensive reform since that time, similar issues apply to the restructured and even new abattoirs.

120 There are two or three small abattoirs in each of the 10 counties in Zhoukou Prefecture in Henan Province that are designed to access mid value markets. Two of the abattoirs in Shenqiu County were visited. Each abattoir slaughtered only about 4,000 cattle per year and produced about 600 tonnes of beef (bone-out). The abattoirs consisted of a small slaughter room with a hook that slides across a ceiling rail, a boning room and a small chilling room hanging 20 half carcasses. The facilities were well tiled, washed and relatively hygienic. There were about 20 workers in the abattoir, who received low wages in this poor, heavily populated area. Most of the workers are employed in the boning room. The labour intensive activity of boning is required by the abattoir to produce 12 different (attractive looking) beef products (cuts) that are individually vacuum packed. Each pack was branded with the name of the customer, which were meat trading companies based in Liaoning, Chongqing, Shanghai and Zhengzhou. The labels included information on the cut, the brand (Chengjian Company) and logo of the certification program used ("Rest Easy Meat" or *fangxinrou* from the Dalian City government). That is, the destination company provided the marketing services while the abattoir provided slaughtering services on contract. The owner of the abattoir was the Party Secretary of the village and the system was established under the World Bank Smallholder Cattle Project. Cattle were sourced from 15 specialised cattle raising villages throughout the county, also established under the World Bank project.

plant in the poor area of Tieling City in Liaoning Prefecture sold its own beef in the mid value range.[121]

These mechanised abattoirs have a number of points in common. First, they are relatively small, are not highly capitalised and have relatively low cost structures. Second, they usually perform the labour intensive slaughter activities of butchering, boning and packaging to reach the demands for mid value markets. Wages in the abattoirs were as low as Rmb500 per month in 2006 in poor regional areas and employment can be highly seasonal and irregular. Third, the small abattoirs usually have cold storage facilities to access distant beef markets. Given that many small abattoirs are located in rural areas close to cattle supplies, there can be long distances to their beef markets.

The budget analysis in Table 7.2 provides further insights into the economic issues faced by small abattoirs targeting mid value markets. The flow of product for this representative abattoir is depicted by the bold block arrows in Figure 7.3, which is drawn from the main study framework. Figure 7.3 also shows the other input–output (cattle purchase and beef sales) options that might be available to the abattoir. Some of these options were run through the abattoir budget and reported as input–output price sensitivities in Table 7.3.

Table 7.2 raises a number of issues. First, costs in the small mechanised abattoirs are low, with cattle purchase cost making up 81% of all costs. Labour costs are low in China, especially in remote areas.[122] Beef marketing costs are significant given that the abattoir is targeting mid value markets that can be a long distance away (although these are lower for service

[121] A mechanised abattoir in Xifeng County in Liaoning had a slaughter capacity of 50 head per day (18,000 per year) but operated well under-capacity because of the bottleneck in the boning room, which could only handle 20 carcasses per day staffed with 16 people and with limited cold storage capacity. The wages at the boning stage and the costs of running the cold room were major cost items (far higher than costs to carcass stage) but these additional boning and chilling stages are necessary to enter into distant and higher value markets. The abattoir had its own butcher shop in Xifeng and agents in Harbin, Shenyang, Dalian and Changchun that operate on an 8% commission. The retail prices of the cuts ranged from Rmb80 for tenderloin fillet but down to Rmb10 for shoulder meat. The design and material (tiles) of the plant would not comply with internationals standards but was nevertheless seemed very clean and well managed for a plant of this size.

[122] For example, a major multinational abattoir (personal communications) estimates that on average (depending on the beef product), labour costs account for about two thirds of all non cattle costs in Australia and about one third of these costs in Brasil. The labour cost component in China is certainly well below that of Brasil. It is also relevant to note that abattoirs are major employers in rural and regional Australia.

slaughtering). Second, while beef is by far the largest revenue item, hides and offal are also major sources of revenue. Third, Table 7.3 shows that the relativities between cattle inputs and beef output prices are a major determinant of profitability.

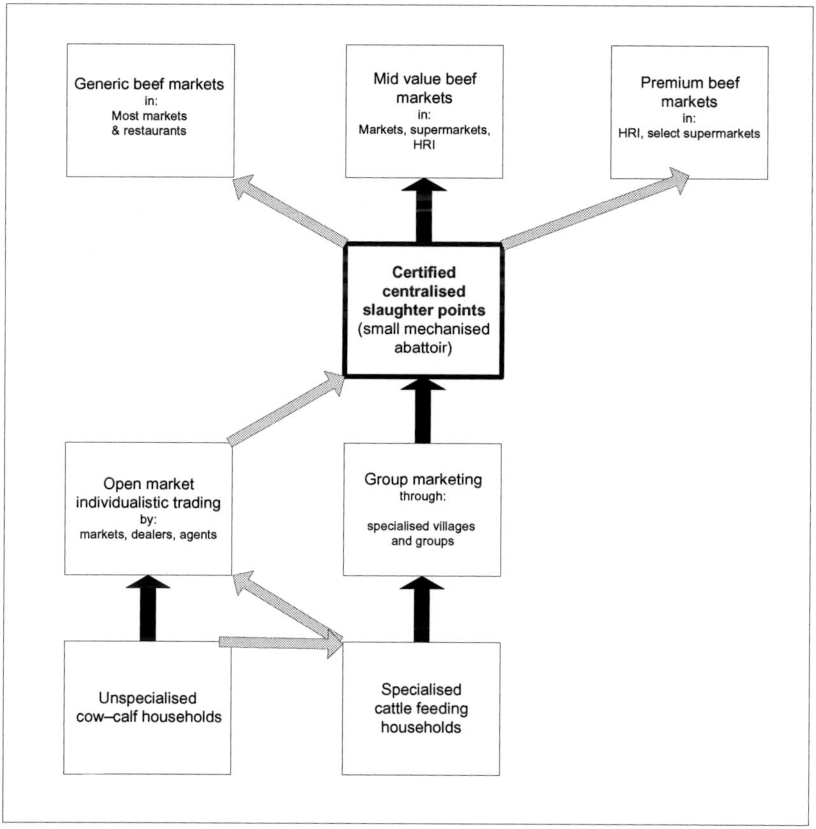

Figure 7.3: Synthesised supply chain for small mechanised abattoirs

Table 7.3 indicates that small mechanised abattoirs are not viable if they sell their beef on to the mass market (with an average beef output price of Rmb15/kg), regardless of the cattle purchased, and clearly can not compete with slaughter households (Section 7.1) in this market. The small abattoirs will be highly profitable if they can sell beef into high value beef markets (Rmb27/kg) but this is an unlikely scenario for a significant amount of their product due to technical, skill and marketing constraints of small abattoirs.

However many small abattoirs are able to sell into mid value beef markets, but not viably if they do so through the purchase of high cost, heavy cattle inputs. It is also important to note in the context of analysing the broader supply chains, that the type of finished cattle that the small mechanised abattoirs buys in to service mid value markets (at the attributes and prices specified in Table 7.3) will be able to be viably supplied by specialised fattening households (see Table 9.3) but will not be able to be viably supplied by feedlots (see Table 9.5). That is, small mechanised abattoirs are most relevant and most closely linked to specialised fattening households.

Table 7.2: Profit & loss statement for representative small mechanised abattoir, 2006

	Rmb/head	Percentage
Revenues	4,780	100%
Beef	3,473	73
Hides	600	13
Edible offal	666	14
Inedible offal	42	1
Costs	4,596	100%
Cattle cost	3,744	81
Cattle marketing	24	1
Inspection & taxes	293	6
Slaughtering	25	1
Boning	50	1
Abattoir overheads	72	2
Beef marketing	388	8
Net return per head slaughtered	184	

Table 7.3: Input–output price sensitivities on net returns to the representative small mechanised abattoir, 2006

Beef sales	Price average all cuts bone-out (Rmb/kg)—value segment		
Cattle purchase	15—low value	21—mid value	27—high value
312kg, Rmb8.7/kg, 52% dressing	–227	**	**
416kg, Rmb9/kg, 53% dressing	–548	184*	**
500kg, Rmb9.3/kg, 54% dressing	–821	85	981
600kg, Rmb9.6/kg, 55% dressing	–1144	–48	1047

All returns are expressed in Rmb net return per head slaughtered.
* Denotes the representative budget results represented in Table 7.2.
** Denotes scenarios where, given specified cattle inputs, it is technically unfeasible to produce specified beef outputs.

In sum, the budget and sensitivity analysis above illustrates that the viability of the abattoir depends to a significant degree on the management, marketing and technical skills of managers to match (cattle) inputs with (beef) outputs and abattoir production systems. Successful small mechanised abattoirs will be skilled in purchasing relatively good cattle, transforming the cattle into relatively good beef through adequate (but not high technology or high cost) infrastructure, and in marketing a significant proportion of the beef products into mid value markets. Reliable food safety and quality assurance systems will have to be in place. This scenario is not unrealistic for many small abattoirs and can therefore be regarded as an appropriate model to supply China's growing mid value beef market.

7.3 Modern mechanised abattoirs

Modern mechanised abattoirs represent the next step up the continuum of scale and modernity. However, the high cost structures of modern abattoirs mean that to be viable they must be able to access premium (high value) beef markets. Because the high value beef markets are not well developed in China (as discussed in Chapter 6) the strong emphasis placed on the development of modern abattoirs by policy makers is largely inappropriate.

7.3.1 Issues in the modern abattoirs sector

China places enormous importance on the development of modern agro-industrial enterprises in all industries from the processing of grain to fruit to dairy. Agro-industrial enterprises are designed to increase levels of industry modernisation and international competitiveness, to foster industry integration, to improve food safety levels, and to increase returns to the producer households that they source inputs from. In addition, local government is interested in developing enterprises that generate within-budget tax revenues that can be retained locally (Oi, 1999). Formal recognition of the importance of dragon head enterprises at central levels appear in a large number of high level documents such as the 2006 Livestock Law, the 10th Five Year Plan,[123] and the agro-industrialisation program.

[123] In the Tenth Five Year Plan (2000 to 2005) the State Economic and Trade Commission stated that "mechanised slaughter should be implemented in all large cities across the country by the end of 2005 and the rate of mechanised slaughter should reach more than 90% in mid and small cities and county towns" (cited in Yu, 2001). This goal has certainly not been reached.

The case of modern abattoirs provides a useful case study of issues that arise in China's attempt to promote agro-industrialisation. As overviewed in this chapter, there are numerous types of abattoirs in China but the ones that are most prioritised are large modern abattoirs that have the capacity to access premium markets and the potential to supply lucrative export markets. As a result, all large beef abattoirs in China are supported by preferential policies and subsidies in one form or another. Examples of specific support programs for larger beef companies are shown in Table 7.4. There are equivalent types of support for smaller beef companies where local government provides land, tax[124] and credit incentives and support for livestock production, especially the organisation of households into associations and the delivery of inputs and extension.

The high level of support afforded to modern abattoirs has a distortionary effect on not only the slaughter sector, but also on other parts of the supply chain. The support attracts companies to take advantage of preferential policies and subsidies, very often without full consideration of underlying commercial forces. These new entrants often do not have a background or experience in the meat or beef industry.[125] These factors, together with the small size of the premium beef market (see Chapter 6), has resulted in many over capitalised and inappropriate investments and overcrowding in the modern slaughter sector. Most modern cattle abattoirs in China would not be viable without support from preferential policies and subsidies, with the possible exceptions of companies like Ke'erqin and Yurun, the former of which reports a return on capital of 5%.

Table 7.4: Preferential treatment given to major beef abattoirs in China

Name of company	Type of support
Hua'an in Hebei	Hua'an, a joint venture established between the (State owned) China Animal Husbandry Group and a German concern was the pioneer of the modern Chinese abattoir sector, treated as a showcase abattoir and has participated in a very large number of programs and projects. The company thrived in the 1980s but is now reported to be in financial difficulty. The subsidiary abattoir that operated for three years in Pingdu in Shandong was closed down, but another Hua'an subsidiary abattoir has reportedly started in Zhanhua (Binzhou Prefecture) in Shandong.

124 For example under the 2006 Livestock Law, land for livestock activities such as feedlots is to be treated as agricultural land for tax purposes rather than industrial land.

125 Examples of companies that have a background in industries like distilling, construction and real estate include SOGREN (Sichuan), Queyuan (Hong Kong), Golden Seed (Anhui), Wuliangye (Sichuan), Southeast Holdings (Shanghai), GITIC (Guangdong), Kangda (Beijing) and Hualing (Xinjiang).

Name of company	Type of support
Fucheng in Hebei	From humble beginnings, the Fucheng abattoir became a success story and was visited many times by the then mayor of Beijing and the then Premier Li Peng, and become a major dragon head enterprise of the beef industry. Majority share of the company was then bought out by the giant Hong Kong conglomerate Wufeng (Ng Fung Hong). The company also operates the monopoly cattle abattoir in Hong Kong and has a major stake (through quota allocation) for both live cattle and beef exports form the mainland to Hong Kong.
Haoyue in Jilin	Haoyue in Jilin has a close family relationship with prominent government officials in the Changchun City government. Haoyue is reported to be the recipient of very large loans (Rmb1.8 billion), but has not been able to service the loans. Without an ongoing line of subsidised financing, it is claimed that the company would not be financially viable. It is also claimed that the relationships influenced Changchun City government regulations that all beef has to be sourced from designated slaughter points, of which Haoyue is the major certified supplier. Indeed, a number of companies have established local monopoly positions through their status as designated slaughter points integrated into city wholesale and retail markets (including for example, Hualing, a range of smaller abattoirs through Xinjiang prefectures and counties, and other cities throughout China).
Dexin in Jilin	The Dexin abattoir in Jilin was established by a range local government partners as a major and ultra modern abattoir funded through a large World Bank loan. It has since changed hands and been declared bankrupt several times and now owned by Haoyue. Amongst the problems attributed to the company is that it originally set cattle purchase prices too high in contracts with producers and did not have sufficient expertise or resources in beef marketing. The company since complained that it could not access enough suitable cattle. Similar problems with cattle supply are also reported to have been experienced by Lingbi in Anhui and Aohuan in Jiangsu.
Caoyuan Xingfa in Inner Mongolia	Caoyuan Xingfa, as a national level dragon head enterprise, did not pay tax on profits, and received enormous support for its role in agricultural development throughout Inner Mongolia and the pastoral region. As reported in Waldron, Brown, Longworth *et al.* (2007), Caoyuan Xingfa was audited over financial irregularities and was declared bankrupt and taken over by a State owned (Chifeng City) enterprise.
Bangjie in Henan	In 2004, Bangjie in Henan had plans to invest Rmb100 million to upgrade cattle slaughter capacity for their "100,000 cattle slaughter project". With the existing 60,000 head, they were to build another 40,000 head slaughter capacity. The funds are underwritten by the Henan Development, the Reform Commission and Henan Agricultural Structural Adjustment Fund (Rmb40 million), and also the State Asset Management Commission (Rmb60 million). The provincial government will make interest payments. Banks are understandably very interested in lending to the company.

Name of company	Type of support
Jinwei Furen in Beijing	Like other abattoirs in Beijing, the Jinwei Livestock Co. is highly subsidised. The Beijing City government has a majority investment in Beijing Jinwei. In turn, the company supplies beef to a range of Beijing government departments and conferences centres. The company took out loans for capital equipment amounting to large interest repayment (Rmb16 million per year) underwritten by town government. Even though this is contrary to World Trade Organisation provisions, it is common in China. Beijing Jinwei has developed links with cattle production bases and small livestock raising areas, including through poverty alleviation programs. The abattoir provides subsidies to households that they are connected to, some of which are in Beijing. The company is also used as a demonstration and research base in collaboration with China Agricultural University Beef Cattle Centre and the China Cattle Industry Association, including in the areas of benchmarking and traceability.
Yuxiangyuan in Beijing	Yuxiangyuan started as a restaurant franchise and branched out into other areas, including cattle slaughter. Yuxiangyuan invested Rmb120 million in establishing its "company + base production area + households" system. From 2001, Yuxiangyuan was designated (by the Beijing Commerce Commission and the Finance Commission) as the only small livestock raising area in Beijing to be a cattle and sheep live storage base. In 2002, the Beijing City Agricultural Bureau designated the areas as the Agricultural Standardisation Demonstration base. It has several times been designated as a demonstration livestock small area and advanced livestock raising small area, and been associated with the "Spark Program". Yuxiangyuan is a designated "Non Public Hazard Foods" (*wugonghai*) production area for beef and has ISO9002 accreditation.
Kangda in Beijing	Kangda is a major designated slaughter point (mainly for sheep but also cattle) for Beijing, where the slaughter is done by households but for which the company receives fees. Kangda received substantial funding and preferential treatment Including poverty alleviation) to establish links with household producers and increase sheep production in the suburbs of Beijing, and to build a very large modern mechanised abattoir (planned to be the biggest in Asia), with a substantial beef line. Construction on the abattoir started but has since been abandoned. For further details see Waldron, Brown, Longworth *et al.* (2007).
Hualing in Xinjiang	Hualing in Urumqi in Xinjiang is set up in a similar way to Kangda above with a designated slaughter point mainly for sheep but also with a substantial cattle slaughter operation. The company receives fees for the slaughtering the households do in their facilities. However, Hualing also takes ownership of a large volume of meat and hides sold through stallholders that operate in retail and wholesale markets owned and managed by the Hualing monopoly. In 2006 Hualing began constructing a very large modern abattoir in Urumqi with a capacity to slaughter 3 million sheep and 300,000 cattle per year with an investment of Rmb300 million. The project is classed as a "special agriculture" project in Xinjiang 11th Five Year Plan (by the agricultural and the industrial divisions of the Development and Reform Commission). This involved at least Rmb1.5 million in grants from Xinjiang government (although this has not been forthcoming yet). The abattoir and meat marketing

Name of company	Type of support
	project will be funded by Hualing itself (which has 13 other much larger subsidiaries in industries such as building material, cars and mining) rather than bank loans. Hualing is prepared to cross subsidise from these other Group activities a in the initial years of operation of the abattoir.
Xisen in Shandong	Xisen in Leling County in Shandong built a large modern abattoir which is integrated with very large feedlot, methane converter, worm farm and other facilities to form a "closed system". This is a philanthropic project of the founder in his home town (now renamed to that of the founder) who become very wealthy through real estate and development projects. Xisen has transformed the town and provided new housing, employment in the company and education services, and established cattle supply contracts throughout the county. The "Xisen model" has caught the attention of Party and government at the full range of levels, for its potential contribution to the beef industry and as a model to modernise rural China. The company has recently (at the end of 2006) changed completely its set of managers (many of which had a background in banking).
Sishui in Shandong	Sishui was established in the late 1990s by the Police with a soft loan from the King of Spain of US$40 million. When national regulations stipulating that police (and the army) departments disinvest their economic interests in 1999, the abattoir was sold. Additional funding (Rmb4 to 5 million) for the abattoir was provided by the Guangdong Development Bank but the company—which by then included a large abattoir and several feedlots—were resumed by county government and now closed.
Xin'ao in Sichuan	Xin'ao in Chengdu was a New Zealand joint venture investment is encountered problems due to inability to secure cattle supplies to standard The local government was also said to overstate figures on cattle supply in the area to attract the investment.
Fucheng Ng Fung Hong, Inner Mongolia Caoyuan Xingfa, Inner Mongolia Nailun, Jilin Haoyue, Nanjing Yurun, Henan Luohe, Chengdu Aoxin, Chongqing Animal Husbandry Group, Henan Shuanghui	On Ministry of Agriculture list of 150 supported agro-industrial enterprises which are eligible for preferential policies, exemption from the payment of profit taxes and the provision of subsidised loans.

7.3.2 Budget analysis of representative modern abattoir

A budget of a representative large modern abattoir presented in Table 7.5 reveals some of the issues confronting the modern slaughter sector.[126] The following are the major determinants of abattoir profitability. First, capacity utilisation impacts significantly on profitability, which is important as nearly all abattoirs in China are operating well below capacity (in the range of 20% to 80%). Second, capital costs can be significant, which is why preferential credit and loan terms are provided to modern abattoirs. As fixed capital costs are incurred regardless of throughput, under-capacity decreases the ability of enterprises to generate revenues to pay the capital costs. To reduce the capital costs of running integrated feedlots and the working capital tied up in the livestock in the feedlots, abattoirs source the vast majority of their cattle through supply contracts or supply arrangements with local government and households. Third, abattoirs are labour intensive and can be major employers in some regions with a major impact on returns.[127] Fourth, the most important determinant of returns to the abattoir is in the cattle input price (which constitutes 70–80% of all costs) relative to beef output price (which constitutes 75–80% of all revenues). These major input–output price relativities are further explored in Table 7.6 which is based on options presented in the synthetic supply chain of the large modern abattoirs depicted in Figure 7.4.

---

[126] Parameters include a slaughter capacity of 70,00 head per year, capacity utilisation of 80%, 40 staff, slaughtering crossbred cattle at 450 kilograms, selling 30% of beef into premium market.

[127] Wage costs for often hundreds of workers vary significantly. In and around Beijing, inspectors are paid by the company about Rmb800–1000 per month, workers in the boning room have wages about Rmb700 per month, and skilled slaughtermen about Rmb1,000 per month. Skill levels of slaughtermen and butchers are important in modern abattoirs. For example, a skilled butcher will have only 15 to 18 kilograms of off cuts from a carcass while an unskilled butcher can have 30 to 35 kilograms. These losses can be worth up to Rmb300. By 2008, the wages of abattoir workers appeared to have escalated significantly to about Rmb2,000 for an eight hour day for eight months of the year (without board and food) in a booming part of Inner Mongolia (Erdos). Part of the increase in cost for the abattoir was associated with the 2007 labour laws which applied extra costs on the employment of temporary workers (insurance, conditions etc.). There is considerable pressure on abattoirs to retain trained slaughter men, as it is common for a whole line of workers and their supervisor to shift their employment to a competitor abattoir at short notice, thus causing major operational disruptions. The labour laws were also said to apply conditions on workers to restrict this sort of short term movement.

Table 7.6 indicates that because of the high cost structures of large mechanised abattoirs, they can not operate viably in the generic low value beef markets (with average prices for all cuts of Rmb15/kg). However, abattoirs are profitable if they can access high value beef markets (Rmb27/kg). If the abattoirs can access high value beef markets, then they are likely to buy 500kg cattle from, and deliver benefits to, specialised fattening households (Table 9.3) which have a competitive advantage over feedlots (Table 9.5) in the supply of this type of cattle. If the abattoir can viably access high value markets, feedlots may also be able to viably supply 600kg slaughter cattle (Table 9.5).

Table 7.5: Profit & loss statement for representative large modern abattoir, 2006

	Rmb/head	Percentage
Revenues	5,703	100%
Beef	4,253	75
Hides	600	11
Edible offal	800	14
Inedible offal	50	1
Costs	5,714	100%
Cattle cost	4,650	81
Cattle marketing	70	1
Inspection & taxes	306	5
Slaughtering	50	1
Boning	70	1
Abattoir overheads	111	2
Beef marketing	457	8
Net Return per head slaughtered	–12	

Modern abattoirs can only operate viably in the mid value beef market (Rmb21/kg) if they are able to source cattle at low prices. However, the lowest priced cattle (312kg, Rmb8.7/kg) will not meet specifications to target mid value markets. It is possible that modern abattoirs may be viable in mid value markets if they source 416kg cattle (for Rmb9/kg), in which case these will be sourced from specialised fattening households rather than feedlots. However, neither specialised households nor feedlots are likely to benefit from the growth of the modern abattoir sector if the modern abattoirs fail to sell beef outputs on to high value markets. The small and overcrowded premium beef market in China (see Chapters 6 and 7) makes it extremely difficult for modern abattoirs in China to sell beef on to premium markets.

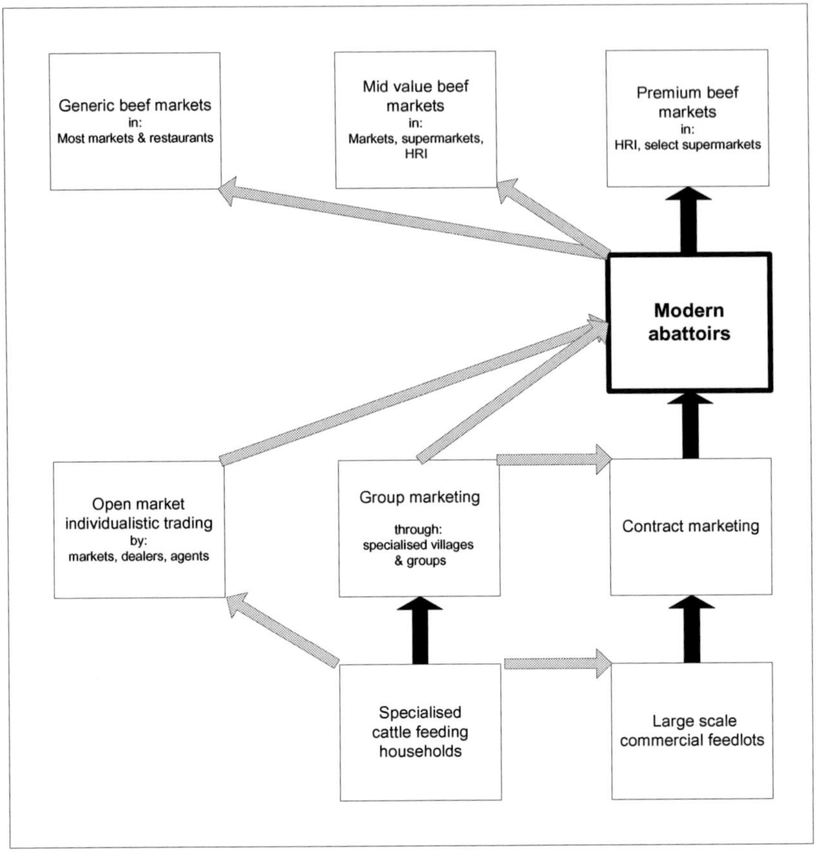

Figure 7.4: Synthesised supply chain for large modern mechanised abattoirs

Table 7.6: Input–output price sensitivities on net returns to representative large modern abattoir, 2006

Beef sales / **Cattle purchase**	Price average all cuts bone-out (Rmb/kg)—value segment		
	15—low value	21—mid value	27—high value
312kg, Rmb8.7/kg, 52% dressing	–365	**	**
416kg, Rmb9/kg, 53% dressing	–641	91	**
500kg, Rmb9.3/kg, 54% dressing	–908	–12*	885
600kg, Rmb9.6/kg, 55% dressing	–1244	–149	947

All returns are expressed in Rmb net return per head slaughtered.
* Denotes the representative budget results represented in Table 7.5.
** Denotes cattle inputs where it is technically infeasible to produce given beef outputs.

The budget and sensitivity analysis results are indicative only but nevertheless help explain developments in the industry in recent times. Most modern abattoirs have struggled to operate viably in China over the 1990s and 2000s.[128] Almost all are reported to be incurring losses and the sector has been highly volatile. As suggested in the analysis, this is most fundamentally due to relativities between (high) cattle input prices and (low) beef outputs prices and, as established in Chapter 6, the small size of the domestic high value beef market.

Difficulty in accessing high value markets, a lack of working capital generated from beef sales, and the pressure for the enterprises to pay back loans on over capitalised facilities puts enormous pressure on abattoirs to cut costs.[129] As cattle inputs are by far the largest cost item (representing up to 80% of all abattoir costs), minimising cattle input costs is crucial. Abattoirs very often do so through opportunistic cattle purchase methods and systems that impact adversely on cattle producers (see Chapters 8 and 9). In addition, attempts to reduce the cost and therefore the quality of cattle inputs mean that many abattoirs can not produce genuine "high quality beef" outputs for high value markets. Under extreme financial pressure to do so, however, many abattoirs will still try to (falsely) market their outputs as "high quality beef". Weak legal (contract) systems and a lack of uptake of quality standards (see Section 6.1.2) mean that such practices are difficult to eliminate.

Another major problem cited by almost all abattoirs is securing sufficient cattle that conform to modern abattoir "high quality beef" specifications, and is widely believed to be another major reason for under-capacity of the sector. It is sometimes argued that in cities such as Changchun and Beijing, abattoirs were designed on a scale too large for the given cattle supply areas which can also be related to policy inducements and distortions.

---

[128] This is in contrast to the 1980s when pioneer companies like Hua'an near Beijing were able to source cheap cattle (Rmb3.6/kg liveweight according to one cattle dealer that used to sell to the company) but was a near monopoly supplier to the expatriate market in China at the time, with technology transfer from the German joint venture partner.

[129] In addition, abattoirs and industry experts report that late or non payment for product by customers (especially beef traders) is in endemic in China. This puts further pressure on highly capitalised plants. The interruptions to cash flow and working capital can be relayed back to cattle suppliers and cattle purchase practices through late or non payment. A consultant who provides accounting services to agricultural enterprises in northern China claims that agricultural enterprises are reluctant to pursue cases of late or non payment through an arbitration commission or the courts for fear of damaging relationships with other customers.

More importantly, however, the inability of abattoirs to secure higher value beef markets means they simply can not afford to pay sufficiently high prices to attract and induce cattle producers to consistently supply cattle to specification. These issues are further discussed in Chapters 8 and 9.

7.4 Concluding comments

Analysis in this chapter examined economic, social and food safety issues facing various categories of slaughter units in China. Based on this analysis it is concluded that China should take more geographically and administratively consistent measures to regulate and restructure unhygienic small scale slaughter households, but in doing so must avoid the creation of local slaughter monopolies, especially if these extend into beef marketing structures. It is also likely that the continuation of preferential policies to develop large modern abattoirs will have detrimental impacts on industry participants in rural areas, because they will be unviable and unsustainable in the absence of a significant high value beef market. Rather, more emphasis should be placed on the development of unmechanised centralised slaughter points and small mechanised abattoirs as a means of developing more market conforming and socially inclusive slaughter sector.

CHAPTER EIGHT

DEVELOPMENT OF THE CATTLE MARKETING SECTOR

This chapter analyses the cattle marketing sector which bridges the demand for cattle by slaughter units (Chapter 7) and the supply of cattle from producers (Chapter 9). Cattle are also traded between different types of producers examined in Chapter 9, including cow–calf households, fattening households and feedlots. Cattle marketing is also closely linked to cattle production (Chapter 9) and to beef marketing (Chapter 6). China's cattle marketing sector has been analysed in detail in Longworth *et al.* (2001, Chapter 8) and Smith (2001).

As a means of disaggregating and analysing China's cattle marketing sector, three types of marketing systems are identified and examined in this chapter. In line with the over arching theme of the study, the three marketing systems are aligned with the value chains to which they are most relevant, as illustrated in Figure 8.1. As the beef industry modernises, the three types of marketing systems will change in relative importance and will evolve in line with the type of product required and the characteristics of buyers and sellers involved in the exchange. The change in relative importance and the evolution of the marketing systems should, in principle, increase the efficiency of exchange, reduce marketing costs, increase pricing accuracy, and increase the capacity of the industry to deliver on consumer preferences. Benefits should accrue across the industry and the full range of industry participants, although the distribution of the benefits will vary among the participants.

The incentives faced by the users of the cattle marketing systems will forge this evolution. In this regard, the sensitivity analysis in Chapters 7 and 9 show that slaughter units and cattle producers have very strong incentives to purchase "good value" cattle at the lowest possible price provided that the attributes of the cattle match their production systems and provide potential for "value adding" to maximise output prices. Larger slaughter and cattle fattening operations also demand large homogeneous lots (or "even lines") of cattle to specification in order to reduce transport and marketing costs and increase production efficiencies. The successful communication and the delivery on these preferences in a production sector comprised of very large numbers of household cattle producers, each raising small numbers of cattle, places major pressures on the cattle marketing systems to evolve.

While the incentives facing market participants provide the impetus for the development of marketing systems, the modernisation of the cattle marketing system will only be sustained if based on well designed and enforced market standards, regulations and institutional arrangements. These issues of market governance were overviewed in Section 2.2.2, discussed in the case of beef marketing in Chapter 6, and permeate the discussion in the case of cattle marketing in this chapter. It is argued that cattle markets, like the Chinese agricultural marketing sector as a whole, is characterised by a lack of enforced or under utilised standards, regulations and procedures. The lack of effective market institutions is a major constraint to the sustained modernisation of the Chinese beef industry.

The two main themes discussed above—the urgency for industry participants to seek favourable input–output price differentials and a lack of effective market governance—means that speculative trading and misconduct occurs more frequently in China than in more developed livestock economies. Thus the discussion below also examines how cattle marketing occurs *in practice* in China and the impacts on cattle producers and other participants that operate in these markets.

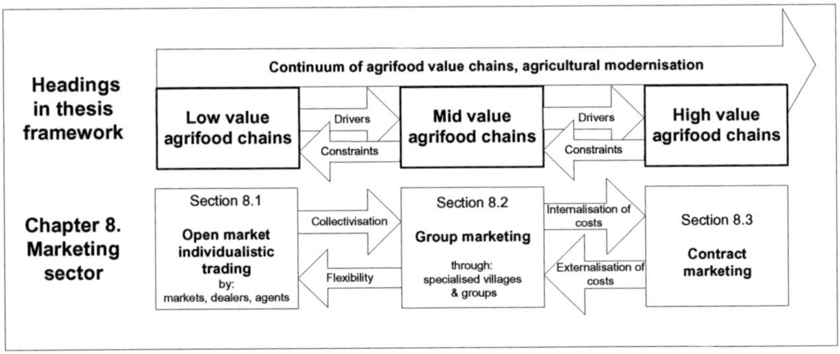

Figure 8.1: Outline of structures, and drivers and constraints to "modernisation" in the marketing sector

This chapter is structured by first examining the so called "individualistic" marketing system (in Section 8.1), which is by far the most important form of agricultural marketing in China. It is argued that this system has generally provided an effective means of improving spatial integration and price discovery especially for bulk, generic and low value products. However, this individualistic marketing system is less effective in terms of generating prices that accurately reflect the true value of more differentiated forms of product in the mid or high value beef chains. In response, attention has turned to the development of alternative marketing systems to service mid and high value supply chains, namely group marketing (examined in Section 8.2) and contract marketing (Section 8.3). The chapter argues that group structures are particularly relevant in servicing mid value markets, and that diversification and increased choice in the types of marketing systems available will benefit cattle producers. In all cases, the potential improvements that could be made in the governance of the marketing systems present challenging but significant and rare opportunities for China to progress agricultural and rural development.

8.1 Individualistic marketing

This section examines the part of China's marketing system that is comprised of an enormous number of individual buyers and sellers, dealers and agents that exchange product through private treaty and spot transactions usually in small volumes per transaction. That is, the system is highly decentralised and fragmented. This system—referred to here as an "individualistic marketing system"—contrasts to that of the central planning era where State marketing agencies dominated agricultural marketing. The dismantling of these agencies in the 1980s led to the rise of the individualistic marketing system which is by far the most important marketing system for the trade of nearly all agricultural products in China. The term "individualistic marketing system" is chosen to differentiate it from the "group" and "contract" marketing systems examined in subsequent sections.

On another definitional note, it is also important to note that "individual" buyers, sellers, dealers and agents usually (but not always) deal with each other in physical market places (rural trade markets of different types). These markets serve as a physical gathering location in which trade takes place through private treaty rather than, for example, systematic lot selling or auctions. While there have been predictions of the demise of rural markets in China (Skinner, 1985), they remain firmly entrenched and are likely to remain so, including in cattle marketing.[130] This is partly because rudimentary physical market places provide efficient forums for the exchange of low value product. However, as discussed below, some

[130] There are large numbers of markets in China that trade cattle. Many of the periodic rural trade (*nongmao*) markets in central China sell cattle in small volumes. Smith (2001) reports that Shanxi has 200 local markets that trade cattle, Anhui has 600, Hebei 1,000 and Henan over 1,500. There are five specialised cattle markets in Shandong that sell up to 50,000 head per year.

larger market places are beginning to introduce basic services to become more relevant to higher value markets, and are therefore also relevant to the discussion of other marketing systems.

8.1.1 Dimensions of transactions in the individualistic trading system

In order to examine in more detail the way the individualistic marketing system operates, Figure 8.2 disaggregates the dimensions of "individualistic" transactions, especially for generic types of cattle and beef in low value supply chains. Based on analysis below, it is argued that the individualistic trading system is a generally efficient and fair channel for the trade of generic product. While information asymmetries exist between sellers and buyers, the asymmetries are limited by the minimal amount of information being exchanged for generic types of products, and there is substantial competition between buyers. In the context of this generally positive assessment, however, qualifications are listed below.

Row 1 of Figure 8.2 shows the industry participants involved in the exchange. Some cattle trading takes place directly between producers (e.g. from unspecialised cow–calf households to specialised cattle fattening households) and processors (e.g. slaughter households). However, much of the trade is conducted through cattle dealers (that take ownership of the cattle) and agents/brokers (that act as intermediaries).[131]

The second row of Figure 8.2 describes the "basis of trade" or the information exchanged in transactions. For the trade of most cattle in China, there are few quality attributes to consider[132] and few downstream preferences to pass back to producers. This simplifies the transaction to within the capacity of a rudimentary and informal marketing system.

In the absence of quality dimensions of the transaction, information used to conduct the trade is limited mainly to quantity (weight) attributes with far less emphasis on quality attributes (such as age and sex).[133] While

---

[131] As slaughter households sometimes need to source cattle supplies from outside their villages or townships, they commission dealers or agents with better knowledge of supplies to buy for them. Many household producers also prefer to sell through agents to avoid direct negotiation and conflict with larger and more experienced buyers.

[132] However, hide size and condition is one of the quality attributes that buyers will consider.

[133] For example, in markets in Shandong in 2006 and in eastern Inner Mongolia (see http://www.eniu.com.cn) per unit prices for cows for slaughter were very similar to those for young bulls for slaughter.

it is easier to estimate quantity than quality characteristics of a product, estimating cattle weights and dressing percentages (or simply the carcass weights of live animals) through subjective or sensory appraisal inevitably entails inaccuracies. These discrepancies have a large impact on buyers and sellers.[134] As suggested in the third row of Figure 8.2, compared to cattle producers, downstream actors such as cattle dealers or slaughter units also have a "sharper eye" in making accurate estimates of dressing percentages and carcass weights and therefore the real value of the beast. This is because they more often see or receive direct feedback on the transformation of cattle into carcasses. At the same time, however, cattle producers are in a better position to know if the cattle have been over watered or fed before sale to increase the weight of the cattle.

While there is potential for some information asymmetries to exist in the open market system, this is largely offset by the competitive nature of markets in most locations. As shown in Row 6 in Figure 8.2, there can be very large numbers of buyers in many parts of eastern China who, as shown in Row 5, buy relatively small quantities of product on a daily basis. As shown in Row 4, payment is made as a spot transaction.

While the sum of the factors above suggest that open markets are generally competitive, as indicated in Row 7, some qualifications apply. There is general agreement amongst industry officials and observers that a degree of active collusion between buyers (slaughter households and dealers) takes place at a local level, based around common business interests and (Hui) ethnic ties. For example, agents claiming to represent sellers will gain a favourable price for buyers in exchange for higher agent fees. Cattle trading is almost always conducted in private between the buyer, seller or agent, with little or no public disclosure or reporting, which inhibits price discovery for other sellers. Cattle deals are usually done through numeric hand to hand signals in secret (under a jacket or in

[134] Traders and agents (justifiably) pride themselves on their ability to appraise live animals by eye. To test their ability, traders at markets in Shandong were asked independently to estimate the weight of particular beast. (This was not a formal experiment verified by weighing on scales, but an informal "test" to establish the scope of the differences in estimates). Liveweight estimates varied by up to 10kg and carcass weights varied by up to 5kg for a 350 kilogram animal, which could be considered a close and skilful guess. However, at Rmb16 per kilogram carcass weight, the error translates into a difference of Rmb80 in the value of an animal, which is a very significant amount given the low per beast returns for cattle producers (see Chapter 9) and particularly for slaughter units when multiplied over many thousands of head per year (see Chapter 7). In both cases, these discrepancies can make the difference between viability and unviability.

private). Dealers are known to tell sellers that they would be given a good price on the understanding that he will not disclose that price to other sellers. Reliable and up to date prices are not listed at market. Against this, however, agents and dealers that operate locally have an incentive to represent sellers fairly in order to preserve their local reputations for ongoing business into the future.

This latter check against anti-competitive trading is less effective in

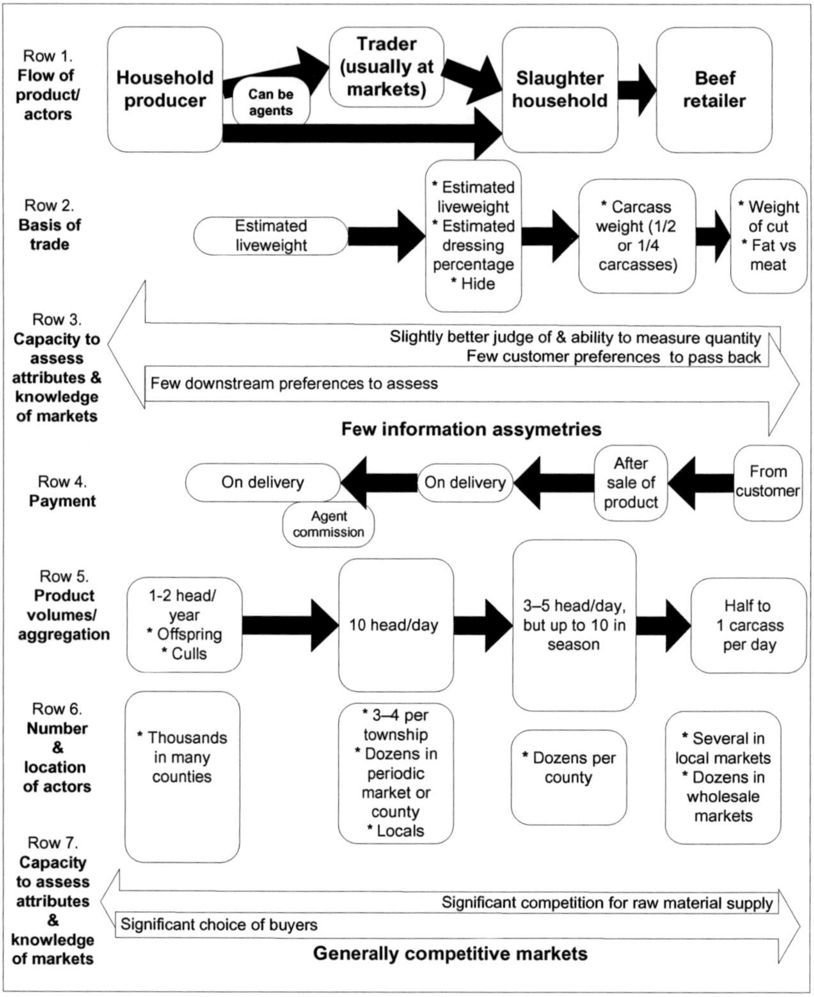

Figure 8.2: Dimensions of cattle and beef marketing for lower value markets

more remote areas of China, where dealers of livestock or livestock products are often itinerant traders, and where trade mainly occurs on-farm rather than in centralised market places. As a result, marketing practices are less competitive in more remote areas of China as described in Waldron, Brown and Zhang (2007, Section 4.3) and Waldron, Brown, Longworth *et al.* (2007, Section 6.3).

8.1.2 Issues in the development of the individualistic marketing system

It was argued above that the individualistic marketing system is a dominant and generally competitive channel for the trade of generic type products in agricultural areas of China. However, it is also argued that the system is largely ineffective in transferring price information (based on downstream preferences) for more specific types of product, and for grade differentials to be reflected in price differentials. As a result, pricing accuracy (or accurately reflecting the true value of the product through price-grade differentials) is "lost" down the supply chain from slaughter units to cattle dealers to cattle producers. As a reflection of this, the individualistic marketing system is unreliable in its capacity to consistently deliver large quantities of specific types of product that consistently meet customer requirements. That is, it is difficult or costly for individual traders to assemble a homogeneous line of, say, 20 head of "good type" of cattle from highly fragmented and small scale producers and markets scattered over large distances. In practical terms, this means that an abattoir in, say, Shandong will find it difficult to assemble, nearly every day of the year, cattle (say, 200 head) to specification (say, 500kg cross bred cattle at two years of age) on the open market, even if the abattoirs buys on physical market places outside Shandong.

As a result of this lack of integration along the supply chain by form of product, abattoirs seek to develop more direct links with producers through group and contract systems discussed below. It is, however, unrealistic to expect these alternative channels to replace the open market system, even for the marketing of higher value agricultural products. For higher value cattle in China, individualistic marketing continues to play a significant role alongside group and contract marketing systems. This is particularly the case for traders and buyers that operate through large inter-regional cattle markets that trade large numbers of cattle over large catchment areas where large numbers of cattle dealers congregate.[135] The scale allows buyers to aggregate lines of cattle or make up shortfalls from other marketing channels. This is important because, as discussed in Section 8.3, abattoirs very often experience irregular supply from contracted producers.

Markets and individual dealers also provide an important and highly accessible sales channel for producers (including groups) to sell higher value animals as an alternative to other channels. Indeed, as discussed in Section 8.2, producer groups usually opt to maintain flexibility in sales channels rather than being tied to direct sales channels such as contracts. This engenders competition and also diversifies risk for producers given endemic problems in Chinese contract marketing systems, as discussed below.

Cattle trading is also a significant source of employment and rural mobility. There are perhaps more than 50,000 cattle traders and agents in China.[136] In addition to these specialised dealers, many households both feed and trade cattle on a speculative basis (see Section 9.2). It is problematic to generalise about returns for these cattle dealers because of highly variable volumes, trading practices and degree of integration (with upstream or downstream sectors).[137] As a guide, however, agents interviewed claimed to make around Rmb20 per transaction, which may equate to Rmb20,000

---

135 Regional cattle markets were visited in Inner Mongolia (Tongliao), nothern Hebei (Weifang), Henan (Zhoukou), Liaoning (Xifeng), Jilin (Donfeng) and Shandong (Yangxin). There are five specialised cattle markets in Shandong that sell up to 50,000 head per year.

136 On the conservative assumptions that every rural township in China has one cattle dealer, there are 150 dealers in 100 cattle intensive counties, 20 dealers in 300 mid sized markets, and 100 dealers in 30 inter-regional cattle markets, there would be 60,000 cattle traders in China. Given China's annual cattle turnoff of 53 million, this would mean a trader transacted an average of 2.5 cattle per day (although direct sales between producers and slaughterers also take place). In some intensive cattle and beef counties in the Central Plains (such as Yangxin County), there are 3,000 to 4,000 cattle and beef traders, and a different cattle agent in every village. Anonymous (2002) reports that in Tongxin County in Ningxia, 42,000 ethnic Hui farmer-brokers (mainly at the Laoheqiao Mutton and Beef Market and also in cashmere trading) generated Rmb3.5 million in sales of agricultural products in 2001, and that the area had reduced rural poverty by 40%.

137 For example: many cattle dealers are also cattle producers; dealers sometimes buy cattle from markets, sell to slaughter households and retain ownership of offal and hides; and while large dealers often have trucks the vast majority rent trucks as required (around Rmb300 for a three tonne truck per day).

per year gross.[138] If this is the case, the incomes of agents and dealers would be far higher than unspecialised cow–calf and average rural incomes in Shandong (Rmb5,000 in 2007) and roughly the equivalent of a staff or wage worker in Shandong in 2007 (All China Data Centre, accessed September 2, 2008).

The future of the individualistic marketing system is inextricably linked with the future of physical markets, and whether they can be upgraded to be a more efficient conduit of higher value and more differentiated products. In this regard, government policies and programs to upgrade larger and more specialised markets play a role. However, incentive structures at local levels are also important. Local government invest in markets partly to extract fees, many of which can be retained at local levels. Fees charged in markets appear to be modest on a per transaction basis but become substantial revenue earners in markets with high turnover volumes.[139] Some local governments and, increasingly, corporate investors also invest in markets to attract trade in the areas of physical infrastructure (weighing, penning, feeding, tie rails, transport and loading facilities and accommodation) and services (such as price reporting). There may be demands for market investors

---

[138] Several agents interviewed in 2006 claim to charge a flat commission of around Rmb20 per head transacted, but can be higher if, for example, slaughter units are in urgent need of cattle, especially at the end of the day where cattle are required for slaughter that night. It is common for agents to transact four head per day. However, trading does not occur every day, and it is assumed in calculations above to take place over two thirds of the year. Unlike dealers, agents do not incur transport, feed costs and other overhead and capital costs.

[139] There are national standards for State fees that can be charged at markets, but there are large local variations in practice. In areas that actively promote the cattle industry, fees were very low or waived. The basis on which fees were paid varies between a flat rate payment, a proportion of the value of the exchange, or a combination. Some of the fees can be retained locally and some transferred to central government. While fees vary, in a large specialised live cattle market in Northern Hebei in 1998 fees included: a market management/construction fee (Rmb8.5 per head, payable to the Industry and Commerce Administration Bureau); a transaction fee (Rmb10 per head, payable to the local government); and disease inspection (Rmb4 per head, payable to the Livestock Disease Inspection Station). Other charges payable to market investors include a service charges for the general use of facilities (e.g. Rmb1 per head), loading facilities (Rmb10 per truck), weighing if done (around Rmb2/head), the penning and feeding of livestock (around Rmb5 per head per night), and meals and accommodation for people. There are also fees for live cattle inspection (of Rmb2–5 per head, payable to the Livestock Disease Inspection Station). The market transacted 60,000 cattle in 1998 with a financial turnover of more than Rmb200 million.

and local government to improve livestock disease inspection services and conduct in order to decrease purchase risks for abattoirs engaged in higher value markets. Some market investors may have incentives to develop more sophisticated services such as auctions, the registration and training of dealers and agents, insurance services, and the possible use of public cattle quality standards. If industry participants (sellers, buyers, dealers, agents) see the value in these more sophisticated services, then markets will continue to play a major role in live cattle trading in China in the future, as is still the case in Australia for example.

8.2 Group marketing

The development of local groups can be seen as facilitating production or processing activities (as discussed in Chapters 7 and 9) but is also inextricably connected with marketing. Local groups include a range of structures ranging from specialised villages to associations, cooperatives and "small livestock raising areas". While some forms of local groups have long existed in Chinese agriculture, the way in which group structures are evolving is examined in this chapter. It is concluded that the development of group structures should be facilitated so that they can play an increasing role in the industry, including mid value agrifood chains, because they have a greater capacity than individualistic systems to produce and sell homogeneous lines of product, and because they allow more flexibility and decision making powers to rural industry participants than is the case in contract marketing systems. However, realising the comparative advantages of group structures requires an improvement in the management capacity of the groups themselves.

8.2.1 Prevalence of local groups

There are mixed reports about the prevalence of group structures in Chinese agriculture. Shen *et al.* (2005) estimate that in 2003 only about 3% of households in China were members of "Farmer Professional Associations", which appears to be a relatively tightly defined form of local group. Livestock was the most highly represented sector, within which cattle (and sheep) were amongst the most highly represented industries in proportion to their size.

It is likely that higher proportions of households are involved in local groups more broadly defined to include specialised villages and towns, associations and cooperatives. Waldron *et al.* (2003a, Sections 5.2 and 3.2.4) discuss these various forms of local structures in the context of the beef industry. One form of local group that has been especially important in the livestock since about 2000 is "small livestock raising small areas" (*yangzhi xiaoqu*) of which there are reported to be more than 60,000 across China. These groups can be highly concentrated in agricultural parts of China[140] and are common in the beef industry.[141]

Observers and policy makers have called for the development of group structures for many years, but these calls have been acted on in recent years. Major central government policies include the Cooperative Law of 2007, the Livestock Law of 2006 and the Number 1 Document of 2005,[142] which

---

140 There are wide variations in the estimates of small livestock raising areas, partly because of differences in definition. The definition of the Animal Husbandry Bureau at the central level is: more than 200 beef or dairy cattle; more than 1,000 meat sheep or goats; more than 500 fine wool sheep; or more than 5,000 chickens. In Shandong, there are reported to be 20,000 small areas for livestock, with a total investment of Rmb3 billion. In Chifeng Prefecture in Inner Mongolia in 2003, more than 30% of all livestock were said to be raised in small areas (including 200 sheep producing small areas comprised of more than 6,000 specialised households) and there are plans to increase this proportion to 70% by 2008 but these plans have not been met, partly because of the demise of Caoyuan Xingfa, the main dragon head company for livestock in Chifeng.

141 For example the Animal Husbandry Bureau of Dezhou Prefecture in Shandong reports that there were 300 small livestock areas for beef cattle in 2006. The criteria is that the areas must have centralised veterinary, disease and feeding systems, and contain at least 10 households. The areas can have as few as 150 head, but as many as 600 with a turnoff of up to 1,200 head per year. Since 2000, Qihe County Animal Husbandry Bureau claimed to have established 270 small cattle raising areas (though with a more lenient definition that that of Dezhou Prefecture).

142 The Number 1 Document for 2005 (titled "Strengthening comprehensive production capacity of agriculture") directed that small livestock raising areas with "good conditions" should be developed through small amounts of credit, financial incentives (in the form of reduced interest payments) and other measures. Central and local governments were asked to increase support for the construction of wastewater infrastructure for small livestock raising areas. The Ministry of Agriculture also selected 20 small areas in seven provinces (Shandong, Shanxi, Heilongjiang, Zhejiang, Henan, Sichuan and Shannxi) for experimental standardised small areas in 2006. The central finance provided Rmb15 million to support wastewater treatment facilities for dairy cattle and pig raising areas (Xinhua Net, 2006).

set the agenda for more local level support.[143] The State is interested in developing local groups for livestock as a means of facilitating: more commercialised, intensive, specialised and modern livestock systems; the control of livestock diseases (which are easier to monitor, vaccinate and treat in group structures); the management of environmental issues (especially effluent treatment and methane conversion); and increased rural incomes.

8.2.2 Benefits of participation in local groups

Local groups provide a number of potential benefits for industry development including marketing aspects of the industry. First, group structures build collective scale of production. Local groups can involve just a dozen households with a total of 100 head of cattle, or many more.[144] In most cases, the members are specialised cattle fattening households analysed in Section 9.2 below.

This concentration of producers facilitates the coordination and delivery of inputs and extension services. Veterinary, breeding and feeding inputs are usually coordinated and delivered through a single organisation either by the group themselves, by contracted providers or local government service providers. Some groups report significant savings from the bulk purchase of inputs and services.[145] In almost all cases, group members and local government also share the cost of infrastructure such as penning and feed treatment facilities. Budget analysis of specialised households in Section 9.2 suggests that savings on these cost items combined

---

143 For example, in 2006, Shandong province planned to provide Rmb2 million for the development of 162 farmer cooperatives in 101 counties throughout the province (Shandong Province Animal Husbandry Office, 2004). In Qihe County, county government organises the services and inputs into small cattle raising areas. Townships organise power and land (0.6 mu average free of contract fees) and grants of about Rmb500–1,000 for shed construction. The subsidies amount to about Rmb10 million for the county as a whole. Households can receive subsidised loans provided they are members of a cattle association.

144 In Shandong Province in 2006 for example: the Atuo township cattle cooperative had 6,400 head of cattle (426 members); the Liuqiao township cooperative had 5,100 head of cattle (270 members); and a village cooperative in Yangxin County had 2,500 head of cattle (280 households).

145 The Atuo cooperative claims to have saved about Rmb10,000 per year in the purchase of Rmb50,000 of veterinary products. Liuqiao claims to be able to buy feed at a 5–10% discount (through government) to reach discounts of Rmb2 million per year.

(inputs and infrastructure) would impact significantly on returns to cattle production.[146]

Perhaps more importantly, inputs can be aligned more closely with production systems to increase technical efficiencies, especially weight gain, which is an important aspect of the profitability of specialised cattle production (see Section 9.2). There are examples of some cooperatives paying for the training of a member in livestock production at provincial agricultural academies, and of employing animal scientists as consultants.

The coordination of inputs and production systems is inextricably connected with the marketing activities of local groups, most significantly because it allows for the production of larger lines of more homogeneous outputs which is attractive to abattoirs buyers. Some cattle cooperatives in Shandong are coordinated across supply chain segments to include both cow–calf producers (that sell calves at 10 months of age) and specialised fattening households. In the case of Liuqiao, the integration extends to a feedlot to finish homogeneous lines of cattle. Liuqiao claims that this allows them to sell to large abattoirs—in particular with a Japanese owned abattoir in Beijing called Jinghe—at higher than average prices. As shown in Sections 9.2 and 9.3, the generation of high output prices have a major impact on profitability.

With regard to sales arrangements, there has been a low uptake in the use of formal supply contracts that stipulate specific output requirements at predetermined prices. Most groups prefer not to enter into tightly specified contracts because of the price risks in a fluctuating market and because of the fluid state of the abattoir sector.[147] Instead, local groups have a general understanding of the product requirements of abattoirs (and their dealers), which keep contact with the groups about the availability of different types of cattle. Finished livestock are then traded at "current market prices", which are also negotiable.

---

[146] If the representative specialised household overviewed in Section 9.2 entered into a group structure, and the costs of feed, feed transport, veterinary services, artificial services and fixed assets (depreciation on pens) were reduced by 10%, the net profits turned off would increase from Rmb121 to Rmb189 per head.

[147] A consultant employed for one of the projects on which this study draws was involved in establishing four cattle production groups in Shandong each with 100 to 200 households that had common inputs, production regimes and produce homogeneous type of cattle. The groups opted against entering into fixed contracts with abattoirs in order maintain flexibility in sales channels (amongst various abattoirs and traders). This is particularly important in China as different abattoirs offer very different sales prices, even for the same type of cattle, due to the wide range of abattoir structures, viability, liquidity of capital to buy cattle, grading and cattle purchase systems, and propensities to negotiate or force low purchase prices.

Group structures offer a range of mechanisms to increase bargaining power for members. For example, the Atuo cooperative in Shandong Province hires a truck to transport cattle to buyers that offer the highest price[148]. There are numerous examples of local groups advertising the availability of cattle, especially on the internet. With the passing of the Cooperative Law of 2007, registered cooperatives can now enter into contracts with legal person status and can independently apply for bank loans, or the cooperative can act as a guarantor on loans by individual household members. Capital flow is an important aspect of livestock marketing because it relieves pressure on households to accept prices on spot markets to meet immediate household costs. The ability of local groups to be able to establish bank accounts and accounting systems so the local groups can send and receive money independently (rather than through government or particular individuals) is important. In addition, there are examples of cattle groups taking out insurance against the death of their livestock, although the insured values were low at about Rmb500 per head.

Local groups in the cattle industry claim that participation in various types of local groups has significantly increased household incomes.[149] Given the potential benefits of participation in local groups discussed above, the modest costs for the households,[150] and the government subsidies, these figures may be realistic for the some of the more successful groups. However, problems that constrain the development of local groups are discussed below.

---

148 This however can be counter productive in remote parts of China. For example a group of sellers from a remote mountainous area of Xinjiang usually sell sheep and cattle to traders on the grasslands in autumn. This limits the number of traders that the households could sell to, and therefore reduces competition, and also enables traders to assess the state of the grasslands and feed resources to know how anxious the households are likely to be to sell the livestock. In an effort to be more proactive in selling the livestock, the group of sellers informally negotiated a sale in Urumqi (many hundreds of kilometres away), assembled livestock from the township, hired trucks and transported the sheep to Urumuqi. The buyers at the market then reneged on the deal, forcing the sellers to house and feed the sheep for many days at high cost. This "holdup" caused a forced sale and the township sellers incurred losses for the households that they sold on behalf of.

149 For example: Yangxin cooperatives reported an increase in household incomes from cattle production of Rmb5,300; Liuqiao by 40%; while officials in Qihe County claim that participation in small cattle raising areas has increased incomes by about 25% to reach Rmb30,000.

150 Membership fees in formal cooperatives in Shandong are usually around Rmb50 per head.

8.2.3 Issues to be addressed in group structures

There are several problems in local groups in the cattle industry. The first is that local groups are very often established with "encouragement" from local officials, especially the Animal Husbandry Bureau coordinating with township government and village leaders. The participation of local officials is most prevalent in the establishment of the local groups which, in order to eligible for approval and funding, is usually done in accordance with government industry plans, infrastructure and technical design standards for infrastructure. In some case the official participation can be continued into the operation of the groups either in subtle ways (for example where the village secretary is the representative of a small livestock raising area) or in more pervasive ways.[151]

Local government is extremely interested in establishing and participating in groups to: conform to higher level edicts to increase the number of agricultural groups; because it facilitates and coordinates their extension and industry development tasks; and because it helps to attract processing enterprises into their region. In regard to the latter point, local government very often act as "brokers" for enterprises to establish local groups in order to capture supply. It is also important to note that many local groups are established and led by particular entrepreneurial households who use the local groups, for example, as a vehicle to sell their breeding material or to capture brokerage fees on the sale of product. The interests of local government, enterprises and entrepreneurs are not always aligned with those of local groups.

It is also very common that once local groups have been established through official support, they then fail to operate as an effective and viable group structure. Many groups exist in name only, are very loosely organised

[151] For example, the Liuqiao Town beef cattle cooperative consists of 10 enterprises including the province livestock and poultry breeding extension centre, the county livestock industry and commerce company, a feedlot, the township veterinary station, the township veterinary network branch, a beef cattle artificial insemination point, the township rural credit cooperative, and the township livestock technology service centre. Resolutions and decisions about the cooperative have been made at a range of meetings—two meetings of the Party Committee, three combined government and Party meetings, and four for all stakeholders in the cooperative. A draft document (containing 6 chapters and 37 clauses) governing the cooperative was presented to the cooperative representatives and passed. These group structures closely resemble the Beef Leading Groups and the local animal Husbandry Bureau companies that were prevalent in the beef industry in the 1990s (see Waldron *et al.*, 2003a, Section 7.3).

and, indeed, are not unlike households acting individually. For example, of the 4,000 specialised livestock cooperatives in Shandong (with 200,000 members), only 1,200 are regarded as "standardised" (Shandong Animal Husbandry Bureau, 2005). The other 2,800 groups are largely dysfunctional and thereby unable to derive the benefits of cooperation discussed above. There can be problems in governance of the groups and it can be difficult to track and allocate costs and revenues for the groups back to individual households within the group.

Increasing the effectiveness and independence of local groups in Chinese agriculture requires a series of measures. Most support for local groups from government and enterprises is for the provision of hard infrastructure and technical support. While important, this should be complemented with more attention on the development of management capacity of the groups. This would include strengthening the ability of groups to assess and choose markets and marketing channels through more complete market information, and by increasing the ability of local groups to provide objective information about their products to potential customers.[152] More proactive activities for local groups include the slaughter of their own cattle (and therefore the marketing of their own beef) through service slaughtering arrangements or direct participation in designated slaughter points operations.

8.3 Contract marketing

It is widely argued both inside and outside of China (see Chapters 2 and 4) that "modern supply chains", vertical integration and contract agriculture will play an increasingly important role in agricultural and rural development in China. If this is going to be the case, a more critical assessment of modern supply chains is required which includes an examination of the real incidence of modern supply chains, and the performance and conduct in these systems. The information provided below suggests that the contract agriculture model operates highly imperfectly in practice in China.

---

152 For live cattle, this includes the breed, weight, age and possibly disease treatment history of the cattle. For commodities such as wool, it requires that the wool is tested for a range of attributes by a recognised authority.

8.3.1 Prevalence of contract marketing

Contract marketing is distinguishable from individualistic marketing discussed above in that transactions are internalised within more formalised structures. Several theories explain the existence of vertical integration and contract marketing. Theories from institutional economics focus on transaction costs, while theories from neoclassical economics revolve around market imperfections such as uncertainty over input supplies, price efficiency, the capacity of the marketing systems to relay the preferences of consumers down the whole supply chain, and market power (Joskow, 2005). Under these conditions, firms seek to internalise transaction costs within vertically integrated structures (Coase, 1937). Significant imperfections and high transaction costs in individualistic and group marketing structures examined in sections 8.1 and 8.2 help explain the significant incidence of vertical integration in Chinese agriculture.

With regard to the incentives of industry participants, downstream actors (processors or retailers) are attracted to short formalised supply chains such as contracts to increase the quantity, quality and consistency of supply. Upstream actors (producers) are attracted to the promise of higher output prices, lower risks and the inputs and services that can be associated with participation in contract systems.[153] Chinese policy makers are interested in the benefits of contract for food safety, disease control and agricultural modernisation objectives.

Agro-industrialisation (*nongye chanyehua*) is widely believed by Chinese policy makers to "suit Chinese conditions" and to be the model on which modern and high value agriculture in China will develop. In Chinese terms, households (*nonghu*) are led (*daidong*) into supply chains (the length of the dragon—*yitiaolong*) by the dragon heads (*longtou*). Contractual relations are predominantly between specialised households and enterprises, although a range of other organisations are involved (such as the model "enterprise + households + groups + association + government" model).

---

153 One reason why contract marketing may not be very prominent in the beef industry is because it provides few benefits to households on the inputs and services side. The beef cattle industry is different to some other livestock industries in China (such as chickens and dairy) where there is a need for more precise and controlled infrastructure, procedures and inputs. In addition, the official extension system in intensive cattle raising areas offers adequate services (such as breeding and veterinary services) at a very low cost, regardless whether the households are contracted or not.

As discussed in Section 2.2.2, Ministry of Agriculture surveys reported in Niu and Xia (2000) and Niu (2002) found that the number of "agro-industrial organisations"—about half of which were enterprises—grew from 11,824 in 1996 to 66,000 in 2000. The number of households integrated with these organisations grew from nearly 20 million in 1996 to 59 million in 2000, or 25 per cent of the total rural households in China. It is important to note, however, that these surveys used a very broad definition of "agro-industrial organisations" to include informal structures and relationships. As these surveys were discontinued and have not been replaced by other comparable surveys, there is little guidance on the growth and incidence of agro-industrial structures in recent years. However, it is entirely possible that China's agro-industrial movement has actually contracted in some agricultural industries over the last decade. This is especially the case for formalised, ongoing contractual supply arrangements between producers and enterprises.

This is certainly the case in the beef industry. Nearly all abattoirs had established supply agreements with household producers at the end of the 1990s. While these are still prevalent in some areas,[154] industry experts agree that household producers have moved away from formal supply arrangements over the 2000s and entered into less formalised arrangements and trading on the open market. As a result, modern abattoirs regularly report difficulties in securing adequate supplies of cattle. To understand why this might be the case, it is necessary to examine the way by which transactions occur in vertically integrated systems.

8.3.2 Dimensions of transactions in contract marketing

Figure 8.3 shows various dimensions of contract marketing in the higher value cattle and beef industry. Row 1 of Figure 8.3 shows the participants involved in contract marketing systems. The most important and direct relationship is between specialised fattening households and/or feedlots as one party (the seller) and abattoirs as the other party (the buyer). Other actors are also indirectly involved, including cow–calf households that sell calves into the system and other downstream actors

---

[154] There can be large numbers of contracted households in intensive cattle and beef production areas. For example, in Yangxin County about one third of cattle raising households were said to be contracted.

such as restaurant and supermarket chains[155]. There are several cases of vertically integrated systems that cover nearly aspects of the supply chain.[156]

Row 2 shows the "basis of trade" (including measurement of product characteristics) that are specified in transactions between the fattening households and abattoirs. Because of the higher product demands, a greater number of product attributes and specificity must be incorporated into transactions than is the case in the trade of more generic cattle on open markets discussed in Section 8.1. Most contracts are based on attributes of the live animal such as breed, age-weight characteristics and an assessment of animal conformation. However, because of the subjectivity and risks associated with purchasing live cattle "on the hoof",[157] some abattoirs purchase on the basis of carcass weight and assessment "over the hooks".[158]

---

155 There are cases of restaurant and in particular hot pot restaurant chains that have direct relationships with producers (e.g. Hongwu, Yuxiangyuan, Caoyuan Xingfa) through slaughter operations of the conglomerated companies. There are no cases of supermarket chains that have direct links with cattle producers. For a relatively low sales volume product like beef—and many other food products—supermarket chains appear to lack the incentives or the capacity to employ the very large numbers of beef and cattle experts, extension staff, legal and financial personnel required to deal with cattle producers directly. Rather, supermarkets simply buy their high value beef from abattoirs that do deal with the households directly.

156 One example is that of Nincheng Group in Dongfangwan Banner in Inner Mongolia which has established a system comprised of: a breeding centre, forage producers (6,000 households), a feed processing company, a feedlot, specialised cow–calf producers (24,250 households), specialised fattening households (350 households), and an abattoir, as well as the city and banner Animal Husbandry Bureaus. Xisen in Lelin County in Shandong had a similar arrangement but also had an integrated methane converter and worm farm to utilise the wastes from cattle production.

157 There are inherent risks in visually assessing for beef quantity and quality characteristics "on the hoof", even for the most experienced buyers. This risk can be compounded by cattle producers over feeding or watering cattle prior to slaughter to increase liveweights. Some cattle producers are known to increase salt in the diet prior to sale to increase water retention and to use various growth promotants. Because of risks in buying cattle on a visual basis, abattoirs will seek to minimise risks by lowering prices (price averaging).

158 "Over the hooks" buying provides a higher level of specificity than visual assessment of live cattle. The carcass can be weighed (without feed, water, offal and other by-products in the live animal) and assessed much more effectively for quality attributes (such as meat colour, fat cover) and disease (inspection of internal organs), including through objective testing. According to industry consultants, there has been an increase in over the hooks buying in the Chinese pig industry in order to reduce risks in pig purchasing.

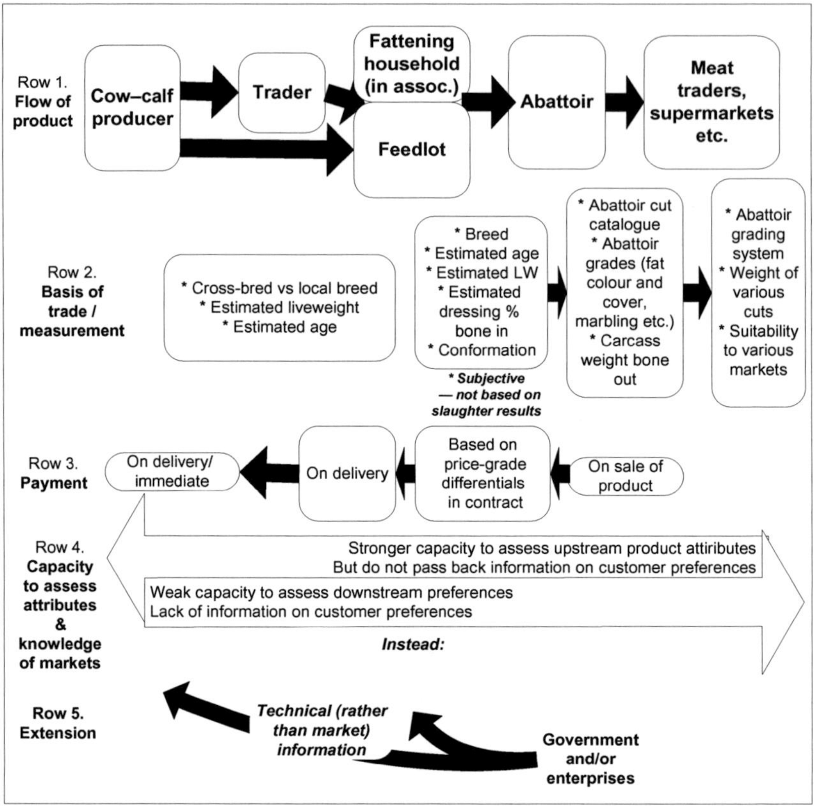

Figure 8.3: Dimensions of cattle and beef marketing for higher value markets

The attributes of live cattle are linked to a pricing scheme, as indicated in the Row 3. Price premiums and discounts are applied to the delivery of the product either in live cattle or carcass form. Thus, price-grade differentials are established in contracts albeit in a simple form.[159] In an attempt to secure

[159] For example, the contract system in Box 2 has two prices for either Grade A or Grade B cattle. Most large abattoirs pay a premium for heavier cattle (carcasses weights over 250 kilograms for example) because of the higher yields of higher value cuts, and discount for lower weights. As incorporated into the sensitivity analysis in Chapters 7 and 9, it is standard for abattoirs to add or discount Rmb0.3 per kilogram liveweight for every percentage point above or below a dressing percentage of 52%. Premiums and discounts are also applied to fat cover and marbling.

supplies, it is also very common for abattoirs to offer a premium over the "market price" at time of sale.

Another dimension of payment is timeliness. It is widely reported that there are delays in payments by abattoirs to cattle producers.[160] This is partly because of the need to assess the cattle or carcasses and process payment to large number of producers.[161] However, many abattoirs also have cash flow shortages and wait for payment from the sale of the beef, which can also be a protracted and risky process as mentioned in Section 7.3.2. The delay in payment is a major disincentive for cattle producers to enter into direct sales arrangements with abattoirs and that can alternatively sell through the open market and receive payment on the spot.

Issues in Row 4 in Figure 8.3 about the relative inability of households to assess cattle—particularly for quality characteristics for higher value markets—compared to abattoirs, are discussed below.

Row 5 in Figure 8.3 indicates that abattoirs often specify the inputs to be used by the fattening households, including breed, feed and veterinary regimes. These specified inputs can be relatively costly and must therefore be offset by relatively higher output prices (cattle). Inputs can be supplied by the enterprises themselves or, more commonly in the beef industry, by local government agencies. Training programs for the contracted households are organised by the abattoirs in some cases. As a result, if households comply with contracts, household production and management regimes are largely determined by enterprises and government, with the main input from households being labour.

8.3.3 Marketing conduct in contract systems

Partial budgeting based on the representative specialised fattening household in Section 9.2 suggests that these households should have strong incentives to enter into contracts with abattoirs. Returns per head for the representative uncontracted household in Table 9.2 (Rmb121) increase dramatically (to Rmb1,434) if the household enters into the contracted specified in Box 2 and if it can produce Grade A cattle. This is

[160] Gong (2007, Chapter 5) found that cattle producers that sell directly to abattoirs wait an average of nearly three weeks for payment.

[161] There are close similarities in this regard to wool sold through the Chinese auction system, where payment delay can be more than three months.

predominantly because of the high output price (Rmb10/kg liveweight) of the heavy finished animal (520kg) stipulated in the contract.[162]

It may be that the simple budget projections suggesting that households have strong incentives to enter into and adhere to contracts do not fully capture real household incentives. However, these apparent incentives are at odds with trends in the cattle contracting sector away from formalised contracts. Abattoirs across China consistently claim that they can not attract enough households to sign contracts and can not therefore source sufficient cattle to specification. Furthermore abattoirs also claim that households readily renege on contracts by selling cattle on the open market before they reach the required finished weight. This occurs if households are in need of immediate capital, cattle prices on the open market increase relative to their (fixed) contract prices, if feed prices increase, and if households are unwilling to adhere to the discipline of a long term feed regime.[163] There are claims that some households fail to adhere to feed regimes, use illegal growth hormones or veterinary products contrary to specification, and that they over-feed or over-water cattle before sale to increase weights, or feed salt to increase water retention.[164]

In addition to speculative, short term behaviour of the households, there may be other underlying reasons why households do not favour contracts. One such reason is that they do not trust the abattoirs. Industry observers

---

[162] Based on contract specifications in Box 2, and assuming the contracted fattening unit buys a 200 kilogram bull calf at 12 months of age (from the cow–calf households examined in Section 9.1), to reach the contract requirements for feeder cattle a daily weight gain of just 330 grams is required. To reach the requirements for slaughter cattle, the daily weight gain is just 600 grams, but is likely to be higher to meet fat cover specifications. The technical requirements to meet these contract specifications are therefore well within the capacities of most specialised households.

[163] The cattle feeding sector is predisposed to this type of speculative activity because there are a large number of alternative market channels for cattle sales, which can be sold in small lines or numbers. This is different to, for example, a typical contracted specialised chicken household that has 10,000 chickens, or a dairy producer that can not readily sell into physical market places.

[164] In this regard, there are some parallels with other livestock industries including households adding sand to wool to increase weight, or adding chemicals to milk to increase the tested protein content. It is often claimed that large dairy companies downgrade (*yaji, yaijia*) their milk purchases, especially when milk prices increase, or when the price of milk on the open market declines relative to that specified with contract producers/milking stations. There are also particular issues in the aggregation of milk as milking stations batch milk from large numbers of households with variable production regimes, meaning that households with higher protein/fat content are penalised as their product is "averaged out" with the milk form other households. There are similarities in the aggregation and testing of wool lots.

claim that abattoirs commonly downgrade product as a means of discounting prices, especially during periods of dropping cattle prices on the open market relative to (fixed) contract prices. As discussed in Section 7.3, abattoirs are under extreme pressure to reduce cattle input costs. The discounting of grades and prices is so common that a specific Chinese term is used throughout the industry (*yaji yajia*), the antonym of receiving high prices for high grades (*gaoji gaojia*). Grades can be discounted by abattoirs on a wide range of attributes including the dressing percentage, fat cover, carcass weight, liveweight, animal conformation or contamination. It is important to note that these attributes are assessed or measured at the abattoir level. Even if slaughter reports on carcasses are passed back to the producers, they are not trusted. Cattle producers are not permitted inside the abattoir to verify the results (for hygiene reasons) and there are limited third party inspection, testing or reporting systems. Downgrading of product has a major impact on producer returns. In the case of the representative contracted fattening household mentioned above, downgrading halves net returns.[165]

In addition to these issues of conduct, there can be a limited number of modern abattoirs within, for example, a city/prefecture for households to enter into contracts with.[166] Even if several are accessible, it is difficult for small households to compare contract terms and purchase practices between abattoirs, which differ enormously and are not fully disclosed. While there are national beef quality standards in China (see Section 6.1.2), and these could be used as a basis to "standardise" cattle purchasing standards, the national standards are rarely used by abattoirs as a basis for cattle purchases.[167] In addition, there are few avenues for legal recourse if contracts are broken or "unstandardised" practices used, especially for individual households. Few households have legal person status to enter into contracts, which are instead signed by other entities on their behalf including local government (townships or service provision stations), local groups (if they are registered as a cooperative), or by companies.

---

[165] For example, if an animal is downgraded from Grade A to a Grade B, per kilogram liveweight prices will reduce from Rmb10 to Rmb9. In addition, the abattoir might report that the dressing percentage is 2% lower than the specified standard of 52%. In this case, budgeting of the specialised fattening household indicates that net profits will reduce from Rmb1,343 (above) to Rmb520.

[166] This is not due to market concentrations for China as a whole which, as noted by Gong (2007, Chapter 4) are extremely low in the Chinese beef industry across the country because of the dispersed nature of the slaughter sector.

[167] For example, in the estimation of carcass weight some abattoirs include while others exclude pelvic fat, heart fat and kidneys, which can account for up to 3% of the carcass weight.

8.3.4 Issues to be addressed in the contract marketing

The conduct of marketing in contract systems is largely a function of developments in beef markets (Chapter 6) and the slaughter sector (Chapter 7). In particular, many modern abattoirs fail to sell a significant proportion of their product on to higher value markets which constitutes only a small part of the overall beef market. Ill-conceived investments, together with the high cost structures of modern abattoirs and overcrowding, means that the majority of modern cattle abattoirs in China are not viable in the absence of preferential policies (see Section 7.3). Their non-viability flows through to the types of premiums they are willing and able to offer producers.

Abattoirs seek to attract and secure cattle supplies through local government and producer groups through the use of contracts on the assumption that they will be able to access higher value beef markets and pass back premiums to producers through higher cattle prices. When abattoirs fail to access high value beef markets, abattoirs then seek to reduce cattle input prices, especially by downgrading the cattle or carcasses. There are a myriad of issues that cause downgrading and reneging on contracts, including late- or non-payment from customers, problems in enterprise governance, or operational problems.[168]

The structure of the abattoir sector is also related to policy distortions. As discussed in Section 7.3, modern abattoirs are supported in policy at all levels of government. Abattoirs are attracted to the preferential policies and subsidies, which can hide the underlying viability of the enterprise. Local officials are usually genuinely and altruistically interested in attracting "dragon head" enterprises because of the perceived benefits in "leading along" the integrated local households. However, there are also strong incentives to attract "dragon heads" at local levels because of the revenues they generate, including fees and taxes that can be retained locally and because the operations provide vehicles through which government officials can claim expenses.

When considering establishing in a given area, abattoirs will assess the availability of local cattle supplies as a major consideration. To attract the enterprises, local government very often overstate the supply of cattle in the area and then go about building supply. This is done through "encouraging" households into the industry, including cow–calf households,

[168] For example, in 1997 and 1998, a very large abattoir was in the process of opening up in Jilin and had already contracted many cattle households. By the time the plant opened many months behind schedule, the cattle were overly fat and weight gains had begun to plateau.

specialised fattening households, local groups, vertically integrated structures and the full range of services including extension, services and infrastructure. The value of these activities is, however, only as strong as the underlying viability of the dragon head abattoirs which, as established in Section 7.3, can be low.

It is widely assumed that producers can benefit from entering into modern supply chains. Indeed, it has been argued (for example Zhou, 2003) that the rural sector will be polarised by the rich households that are integrated into these systems and the poor households that are not and sell on the "open market", thus creating a "dual economy". Analysis in this chapter suggests that the main issue is not just one of *if* the household is engaged in the modern supply chains, but *how* and what type of supply chain with which it is engaged. The conduct of the contract parties and the policy, legal and institutional environment in which the contract is executed is paramount. In this regard, the development and administration of effective standards, grading systems, market information and legal systems is required, as discussed in Waldron *et al.* (2003a, Chapter 11).

The combination of these factors raises the critical issue of market access for producers. It was established in Section 7.3 that large modern "dragon head" abattoirs unable to access high value markets for a sufficient proportion of its outputs will be unviable. These abattoirs do not therefore provide a sustainable vehicle for cattle producers to access high value markets. Indeed given the small size and the low growth of the premium beef market in China, cattle producers are unlikely to gain from establishing contracts with large modern dragon head abattoirs.

Box 2. Example of an abattoir-household contract in Shandong

Background information on the contract system

The company claimed to have established contracts with 10,300 households at the end of 2005. In addition to the prices written into the contract, the company provides a range of other incentives and services to the households. For example:

- The company provides training to the households
- Provides feeder cattle to households on a 2:1 basis (if the household raises two feeders, the company will provide another)
- Will provide cows on a 1:1 basis (if the household has one cow, the company will provide another).

- If the contracted household can produce an animal that meets Grade A criteria (10 mm fat cover and marbling) then the company will add another Rmb600 to Rmb1000 to the purchase price and another Rmb400 to Rmb500 for a Grade B animal.
- There are also measures to help the household integrate cropping, cattle and worm production. These incentives are designed to build cattle numbers in the county, which has only 30,000 head turnoff (from 60,000 head in stock).

In this particular case, the live cattle subsidiary company (of the larger group company) purchased cattle as light as 300 to 400 kilograms liveweight. Some of these cattle are slaughtered and some are finished in the very large feedlot of the subsidiary company to a weight of about 550 kilograms. The feedlot turned off about 20,000 of cattle in 2005, which makes up one third of the 60,000 annual slaughter of the abattoir.

Contract example
Title of contract: "Production and purchase cooperative agreement"

Contract party 1: _____ company [name deleted, the cattle purchasing arm of the abattoir]
Contract Party 2: Household producer

To develop the local economy, to build a mutual gain for the household and the company, both parties will coordinate to enter into the cattle production and procurement agreement with mutual trust:

A. Party A [the purchasing party] has responsibilities and rights. The purchaser:
- Is obliged to buy the cattle that are to standard. The purchase price will be Rmb0.2 higher than market price. The floor price for A grade cattle is Rmb10 per kg and Rmb9.6/kg liveweight for B grade cattle.
- Will provide cattle production standards and requirements, disease prevention and veterinary products, cattle production management, worm raising skills and leadership, and will buy all the worms and worm casings (worms are also raised on the cattle manure produced by the household)
- The agreement can be terminated if an uncontrollable disease occurs within the administrative area

B. Party B [the households/contracted producer] has the responsibilities and rights to:
- Use their own money to raise [____ number of] cattle with [____ number] identification numbers and raise the cattle to company specifications (all new cattle must be marked and recognised through signatures by both parties)

- Must provide to the company all production information and follow company production regimes
- All cattle to specification must be sold to the company, or if the cattle do not reach specification can be sold to outside parties

C. Purchase standards (see attachment)
D. Cooperation period [asked to fill in for specific start date to a specific finish date]. The period can be extended is agreed by both parties.
E. Both parties understand the terms of the agreement
F. There are 2 copies—for Party A and B

Dates and signatures of both parties

Example of cattle purchase standards

___________________________________ [name of company, name deleted]

Purchase standards

1. Feeder cattle purchase standards

- Basic requirements. Free of disease, recently consumed feed, water and urine has been expelled, and the rumen operating normally.
- Breeds. Pure breeds: Luxi yellow, Qingchuan yellow and Nanyang yellow etc. Crosses: Luxi females crossed with Limousin, Simmental, Charolais, German Yellow sires.
- Sex, age and weight. Steers above 350 kilograms, two pairs of permanent teeth or less. Bulls above 300 kilograms, one pair of permanent teeth or less.
- Conformation. Normal appearance, rectangular or cylindrical in shape, long torso, good muscle development
- Head and hair. Normal characteristics for the breed, shiny, soft and elastic, unmarked
- Head area. Wide forehead, thick face, head a bit short, big and spirited eyes, mouth big and wide, neck short and thick
- Front quarter. Shoulders normal and full, chest circumference big, wide and deep.
- Mid area. Back is wide, flat and straight, long rib bones, flank tight not loose.
- Hind quarter. Hindquarter even, wide and rounded, buttocks long, flat and wide, distinct tail base.

- Legs and feet. Strong legs, distinct joints, good leg strength, normal pacing, feet/hooves big and round
- Fat score/body fat of 5 or above
- Temperament. Eyes alert, ear and tail movement

2. Fattened/finished cattle

- Breeds. Pure breeds: Luxi yellow, Qingchuan yellow and Nanyang yellow cattle. Crosses: Luxi females crossed with Limousins, Simmentals, Charolais, German Yellow sires.
- Age and sex. Three pairs of permanent teeth or less. Bulls: two pairs of permanent teeth or less.
- Heavier than 520 kilograms
- Body conformation. Free of diseases, hide unmarked, must have distinctive beef cattle characteristics
- Fat. Usually fat score of eight or above, fat thickness under back area will be about 10mm for a pure breed and about 5mm for a cross breed animal.

8.4 Concluding comments

The detailed examination in this chapter of the way that cattle marketing takes place in practice in China, reveals several themes that have implications for policy and theory. It is widely believed that agricultural markets in China, which are dominated by "individualistic" marketing structures, are efficient and competitive. Analysis in this chapter conforms to this finding in the case of the trade of low value generic products. However, the modernisation and segmentation of industries in China requires the development of more sophisticated marketing structures and regulations. Attempts in China to do so through the development of contract agriculture have in the beef industry—and many other industries—have been largely unsuccessful due to the underlying unviability of the contractors (large modern abattoirs) and poor governance of the contract systems. While policy responses to the problems may see contract agriculture play an important role in agricultural marketing in China in the future, these measures must be undertaken in parallel with the development of other marketing channels. The development of local group marketing structures, the development of more sophisticated services at physical markets and the adoption of basic grades and standards in trade will become increasingly important in developing marketing structures that effectively service mid value beef markets in China.

CHAPTER NINE

DEVELOPMENT OF THE PRODUCTION SECTOR

Production structures in Chinese agriculture are dominated by small scale producers but there has been significant change in recent years including the development of specialised household and large scale commercial production structures as depicted in Figure 9.1. The economic, policy and social forces behind this change are discussed in this chapter in the context of cattle production. The analysis draws on budget and sensitivity analysis as described in Box 3 and Box 5. The analysis suggests that unspecialised cow–calf households are unviable in commercial terms, while in the cattle feeding sector, specialised households represent a more viable and flexible form of production than feedlots.

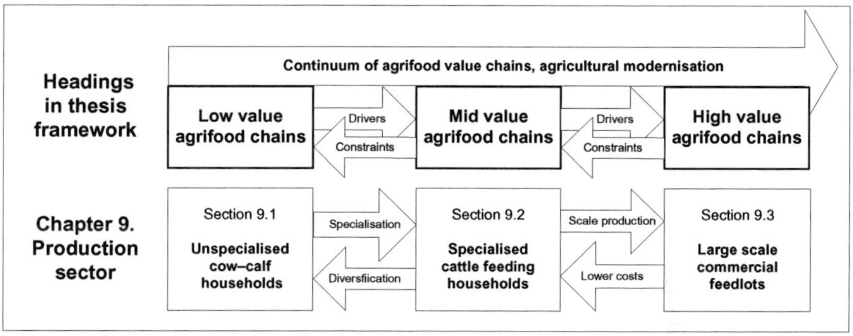

Figure 9.1: Outline of structures, and drivers and constraints to "modernisation" in the production sector

9.1 Unspecialised cow–calf households

Unspecialised households are, by definition, diversified across a range of activities.[169] A related characteristic is that they are semi-subsistent, where a significant proportion of agricultural outputs are consumed on-farm or

[169] The Ministry of Agriculture formally defines specialised households as those that devote 60% of their resources toward a particular activity, although in practice output volumes (in the case of cattle, in stock or turnoff) is used.

are used as inputs into other on-farm activities (transferred) without full reference to off-farm markets. Production inputs such as labour for some members of the household are valued at low opportunity costs.

Small scale, unspecialised and semi-subsistent production systems remain predominant in Chinese agriculture, including the cattle industry. In agricultural areas of China, unspecialised households have traditionally held one or two cows primarily for draught purposes. Cows intermittently produce a calf, which is fed with crop by-products in cut and carry systems or grazed on collective areas and roadsides and supplemented with a few handfuls of grain produced by the household. For a comprehensive description of unspecialised cow–calf production systems in China see Longworth *et al.* (2001, Chapter 5). There were about 13 million of these households in China in 2005 and some cattle intensive counties can have around 100,000 unspecialised cattle producing households. These small scale unspecialised households still dominate the cattle production sector in terms the number of producers and proportion of cattle turned off, as shown in Table 5.2.

9.1.1 Issues in unspecialised agricultural household systems

Despite the dominance of these small unspecialised households, Chinese agricultural structures are undergoing commercialisation and specialisation as reflected in a number of indicators. First, the proportion of agricultural product that is bought and sold off-farm as opposed to self produced and consumed is increasing (Gale, 2005) to reach nearly 100% for products such as fruits, nuts and vegetables (Huang *et al.*, 2007). The proportion of cattle (and sheep) killed for self consumption in Inner Mongolia is also low and decreasing (see Figure 5.11). Second, technical indicators such as livestock turnoff rates signal the proportion of livestock that are sold off-farm. In the case of cattle, the proportion of China's cattle herd being turned off for sale and slaughter is increasing compared to the number being kept on-farm in inefficient production systems or for draught purposes, as shown in Figure 5.9 and Figure 5.10. Third, data on the scale of production also indicate change in the relative importance of producer categories. The scale of production in the beef industry is increasing as shown in Table 5.2, indicating a diminishing role for small scale unspecialised household producers. The reasons for these trends are explained below.

9.1.2 Budget analysis of unspecialised cow–calf producers

To understand the issues facing unspecialised household producers, a representative budget for an unspecialised cow–calf household was developed, as reported in Longworth *et al.* (2001, Chapter 5) and summarised in Box 3. The discussion that follows is primarily concerned with examining the change that has occurred in the unspecialised cow–calf household production sector in recent times. To this end, Table 9.1 presents budget results based on parameters from 1998 in the Central Plains. The budget is then run again using parameters from 2006.

The budget results for 1998 indicate that unspecialised cow–calf production *decreased* returns to labour, capital and management by Rmb229 (as shown in Table 9.1). The negative returns occur because revenues (from calf sales, culled cows, draught and manure) were lower than costs (especially feed). This was particularly the case as calves were kept on to slaughter age in inefficient feeding systems. These budget results conform to anecdotes that some households do not use their draught cows for breeding and that calves are sometimes killed at birth to avoid the costs of feeding them.

Box 3. Background to the household budget analysis

Detailed simulation models of household cattle producers were developed to construct the budgets reported in Longworth *et al.* (2001, Chapters 5 and 6). Activities in the budget analysis include the production of various crops, various livestock types including cattle, and also off-farm income generation. Costs and revenues are attributed to each of these activities to form synthetic household budgets. At the core of the simulation models is a reconciliation of household production and use of feed, labour, manure and other inputs, as well as household outputs. The models are also designed to disentangle, or clearly identify, the opportunity costs of various intermediate inputs such as feed and manure.

The models are based on a steady state (year in, year out) production system and so cannot capture the development or contraction phases of households moving into, or out of, cattle production. Thus, to understand more fully the incentives facing households not currently engaged in cattle raising, or the opportunity costs for cattle households considering moving out of the industry, the findings of the steady state models need to be augmented by these adjustment costs.

Data used to construct the models came from survey questionnaires and detailed interviews with cattle raising households in the Central Plains provinces of Shandong, Henan and Anhui, and the Northeastern provinces of Liaoning and Jilin. A total of 26 unspecialised cattle-raising households and 18 specialised households completed valid survey forms. The information contained in these surveys and follow-up interviews, augmented by other secondary information, was used to develop models of representative cow–calf households and representative fattening households. The surveys were conducted in 1998 when cattle and feed prices had moderated and were relatively stable compared to the fluctuations of preceding years.

Longworth *et al.* (2001, Chapters 5 and 6) provide a detailed examination of the budgets, including the disaggregation of revenue and cost items. Returns to different factors of production, and for different valuations of intermediate inputs and difficult-to-value outputs, are calculated to reveal more about household incentives from a variety of perspectives. Household revenues, costs and returns are also disaggregated among crops, other livestock, cattle and non-farm income to highlight the relative importance and opportunity costs of these different groups of activities.

The budgets presented in Table 9.1 for the unspecialised household and in Table 9.2 for the specialised household are summaries of the more detailed examination in Longworth *et al.* (2001). That is, the tables present a summary of the budget results only and exclude many items (such as off-farm labour, own consumption of on-farm produce, costs such as depreciation and revenues such as manure) which means that the columns and rows will not aggregate to the equate with the totals.

In the case of specialised fattening households, the model is used to test the sensitivity of net household returns to (feeder cattle) inputs and (finished cattle) outputs. This form of sensitivity analysis is chosen because these cattle inputs and outputs constitute by far the largest revenue and costs items, and therefore have the greatest impact on household returns. The prices and the attributes (weight, age, days on feed) of the cattle outputs of the specialised households are also consistent with the cattle inputs of small abattoirs (Table 7.3), large modern abattoirs (Table 7.6) and feedlots (Table 9.5). The integration allows for an assessment of the markets that the specialised households will be selling different type/priced cattle into. This helps establish relationships between value chains. While other factors that impact on household returns (such as feed conversion and feed prices) also impact on returns, it is not been possible in the confines of this chapter to report in detail sensitivity analyses on these factors. However, these factors are briefly discussed in the text in the chapter and in detail in Longworth *et al.* (2001, Chapters 4 and 5).

Table 9.1: Summary of revenues, costs and net returns for unspecialised cow–calf households in agricultural areas, 1998

	Farm				Off-farm
	Cattle	Other livestock	Cropping	Total farm	
Total revenues (Rmb)	**1,406**	2,254	12,096	15,756	4,025
Calves and culled cows	1,000				
Manure & draught	405	54			
Own feed			2,659		
Total costs (Rmb)	**1,635**	1,870	4,331	7,837	
Own feed	1,312	1,348			
Manure & draught			459		
Returns to labour, management & capital (household returns)	**–229**	384	7,765	7,919	4,025
Returns to management	**–898**	–691	–95	–1,687	

Parameters: The household held one cow that produces 0.6 of a calf, sold out at 24 months of age with a daily weight gain of 300 grams and a straw based diet supplemented by small amounts of grain.

The negative budget result raises the question of why cattle production in unspecialised households increased dramatically throughout the 1990s. One reason is that strong official pressure during this period and a series of industry programs including the Straw for Beef program (see Waldron *et al.*, 2003a, Chapter 9) pushed households into the small scale cattle production. As this pressure and these programs were wound back, the number of unspecialised households engaged in cattle production also fell. Another reason for the increase in unspecialised cattle production in the 1990s is that the unspecialised households undervalued, or did not value at all, items produced and consumed on-farm. That is, items produced and consumed on-farm were not valued using market based opportunity costs. The non- or under-valuation of these items gave unspecialised households the impression that cow–calf production was a worthwhile activity. To support this explanation, if the key items produced or consumed on-farm (feed, draught and manure) are excluded from budget calculations, then returns to labour, management and capital from cattle production are positive at Rmb677. If feed is valued at the costs of production, then cattle production is a break even activity.

To understand how the economics of unspecialised cow–calf production might have changed since 1998, the budget exercise was repeated using major parameters gathered on fieldwork in western Shandong in 2006.[170] Most notably, many cow–calf households in eastern China reduced the feeding period and turnoff age of the calves, producing younger, lighter feeders to sell to specialised fattening households. At the same time, however, many households continued raising calves to slaughter age (more than 300 kilograms) to sell to slaughter households, usually through markets. The sales channel options for unspecialised households in shown in Figure 9.2.

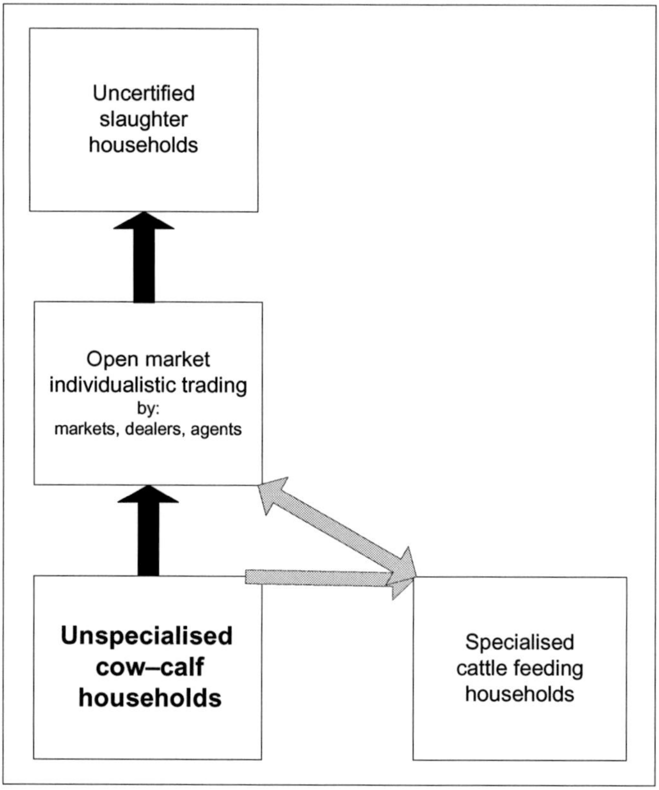

Figure 9.2: Sales channel options for unspecialised cow–calf households

[170] Feed prices were similar in 1996 and 2006, calf prices increased (+25%), technical efficiencies increased (feed conversion ratio from 6 to 5.35) and there has been a change in production systems (3 cows, 100% calving rate, calves sold out at 12 months at 200kg liveweight).

If the cow–calf household sells young feeders to specialised cattle fattening households to reduce the feeding period in the cow–calf household, and if inputs and outputs in the household system are valued on market based opportunity costs, then in 2006 "returns to labour, capital and management" are marginally positive at Rmb75 per cow held. However, households in the in Central Plains areas of China also have increased access to other agricultural and off-farm opportunities and may, therefore, include the costs of their own labour when assessing the value of raising cattle.[171] In this case, the "returns to capital and management" are negative at Rmb–523 per cow held.

This simple representative budget analysis explains several major changes in the industry. First, there has been a reduction in the number of small households that raise cattle in places like Shandong, as shown in Table 5.2. Between 2003 and 2005, the number of producers in this category in Shandong (around 1.2 million) decreased slightly but in the context of an increase in total turnoff of 600,000 head. Consequently, the average turnoff in this category increased from 2.4 to 3 head. Second, there has been a plateau in the number of breeding females in China and in some Central Plains provinces in particular as discussed in Section 5.4.2. Third, there has been a shift in cow–calf production from more developed Central Plains areas (such as Shandong) to more heavily populated and less developed parts of the Central Plains (such as Henan), or remote provinces outside the Central Plains (such as Hubei and Jiangxi). This conforms to widely held impressions in government and industry that only poor households in poor areas have incentives to raise cows and calves.[172]

Structural change in the industry, however, has not occurred at the breakneck pace suggested by growth in incomes, worker migration and rural industrialisation in Central Plains provinces like Shandong, Hebei and Anhui. Large numbers of households continue to engage in small scale cow–calf production despite the apparent unprofitability of the activity if market valuations are used. This is likely to be because while off-farm and urban work opportunities have been taken up by working age members of the household, older and younger generations remain on-farm to feed the cattle and manage the family farm, in which cattle still play an important role. Thus, the transition from semi-subsistent to commercial agriculture

---

171 Annual average wages in surveyed areas in Shandong were reported to increase from around Rmb4,000 in 1998 to Rmb6,000 in 2005.

172 For example, after a tour of Henan, Shandong and Anhui in 2005, the Beijing Animal Husbandry Bureau estimated that areas with annual per capita incomes of more than Rmb1,500 would not have an economic incentive to raise cattle.

may take the passing of the older generation to constitute a transformation of the unspecialised cow–calf sector in relatively developed agricultural areas. As this happens, the more inland and less developed parts of China may become more important in the cow–calf production sector, a prospect raised for the broader livestock sector by Rae (2008) and Rae and Zhang (2009).

Another reason why many millions of households in eastern provinces of China may still raise cattle is that there are substantial income differences within regions including villages (see for example Xing *et al.*, 2006). These incomes differentials have led to the development of innovative relationships between households, including in cow–calf production, as discussed in Box 4.

Box 4. Credit arrangements in a cattle intensive village in Shandong

The economy of Weijiazhaohang Village in Sanwan Township in Yangxin County in Shandong revolves around cattle. Next to the houses in the village is a large tract of land that has been dedicated to cattle, which dominate the sights and smell of the village. There are 198 households in the village that have an average of five head each. Between 1998 and 2002, households in Weijiazhaohang Village sold off breeding females as households left the cattle production sector for other work. Households and local officials believe that it is still difficult to earn a profit from cow–calf production and that margins in fattening sector have also become tight. However, they also claim that the profitability of cattle production started improving from 2002 and especially in 2004. Female cattle numbers started increasing again in Weijiazhaohang village in 2005. They claimed that one cow can now earn the household Rmb400 to Rmb500 but this excludes the costs of the household labour. It was claimed that a fattening household could make Rmb100 per month per head if they owned the animal, but less (Rmb75) if it was done on contract because they lost the 25% of the increase in value mentioned below. The area had developed some innovative arrangements for cattle production, under two main models:

- Under the first model, the household receives cattle from an abattoir in Yangxin County called Zaiwangcheng. The abattoir gives the feeders to the households at about 400kg and records the weight, price and value of the cattle. For example at Rmb10 per kilogram liveweight, the cattle would be worth Rmb4,000. Then the household feeds the livestock to about 500kg liveweight which, at a market price of Rmb12 per kilogram liveweight, is worth Rmb6,000. The increase in value is therefore Rmb2,000. 25% of this increase in value (Rmb500) plus the purchase price of the feeder (Rmb4,000) is deducted from value of the finished animal (Rmb6,000), leaving the household with Rmb1,500 in revenue. The household of course has feed and other costs to deduct, but estimate they can make a profit of about Rmb600 per head for fattening the cattle. However, as suggested by the crude calculations above, the profitability is highly sensitive to markets prices. For example, if the purchase price and the sale price are the same (Rmb10), then the profitability of cattle fattening for the households is marginal or negative.

- Another model is where one household (which is capital poor but feed and labour rich) fattens cattle for another household (which has the capital to invest in cattle but is feed and labour poor, often because they are working off-farm). The relationship between the households is similar to that outlined above for the abattoir. On sale of the finished cattle, the fattening households is obliged to give first offer of sale to the household that provides the cattle but if it declines to buy the cattle at an agreed price, the fattening household can sell the finished cattle (through any channel, including markets or direct to abattoirs) and then return the money to the household that provided the feeder. In the short time that the model has been operating, there were said to be few problems because both types of households wanted to maintain the relationship.

9.2 Specialised cattle fattening households

Compared to unspecialised households, specialised households tend to have a larger scale of production and are more integrated with markets, including purchasing items such as feed into the household systems for cattle fattening. While there are cases of specialised cow–calf producers, most specialised cattle households tend to be in the cattle fattening sector. Chinese production systems are undergoing a gradual transition from being totally dominated by small scale, semi-subsistent household producers toward the inclusion of larger and more specialised households. From a very small base in the 1980s, household specialisation (*zhuanyehua*) has become a significant movement in Chinese agriculture, as indicated by official scale or production data and various surveys.[173]

Scale of production data in Table 5.2 reports that there are around half a million households in China that turnoff between 10 and 100 cattle per year, making up about 27% of China's cattle turnoff, up from 21% in 2003

[173] A survey by the Centre for Chinese Agricultural Policy asked village leaders a yes/no question of "Do farmers in your village specialise in the production of a field crop, tree crop or livestock commodity"? The percentage of "yes" responses increased from 21% in 1995 to 30% in 2004 (cited in Huang *et al.*, 2007).

(see Table 5.3). These proportions are not dissimilar to those in the pig and sheep/goat industries though scales of production are much higher in the layer and broiler industries.

Household specialisation occurs as unspecialised households move out of diversified small scale, semi-subsistent activities and scale up a particular activity, either on-farm or off-farm. Up to a million households exit the unspecialised cattle production sector every year, of which tens of thousands of previously unspecialised cow–calf households "step up" to become specialised fattening operations, sometimes as a step toward moving further up the industry supply chain into feedlotting and cattle trading in flexible, speculative, entrepreneurial operations. There have also been many cases on fieldwork where cattle fattening households moved off-farm into cattle slaughter and beef trading operations in family and group networks. In many of the specialised households visited, younger members of the household ran the specialised operations and had chosen to either remain on, or return to, the family farm from education and work in other areas. Mobilisation channels within an industry structure provide important micro-level insights into the broader process of rural development in China.

Another major aspect of specialisation at the individual household level is that it is closely connected with the development of group production and marketing systems. Indeed, being a specialised household is almost a prerequisite to participating in group systems including specialised villages, cooperatives, associations and "small livestock raising areas" discussed in Section 8.2. In turn, these local level groups amalgamate to build county and regional specialisation. Relatively centralised and intensive production structures facilitate breed improvement, livestock disease treatment, feed (especially straw) treatment and effluent control (through collection and methane converters). Specialised households are also closely tied with the development of modern supply chains. Processing enterprises—such as the modern abattoirs—source much of their "raw material" inputs from specialised households (or feedlots discussed below). They do not deal *directly* with unspecialised households.

9.2.1 Issues in specialised agricultural household systems

Specialised activities have the potential to generate higher household incomes through cost economies, scale or technical efficiencies and by selling into mid and sometimes even high value markets. However, household specialisation also increase risk profiles for the households as profitability depends heavily on price alignments. Compared to unspecialised households, specialised households are closely linked to markets and exposed to price cycles which have been severe in the Chinese livestock industry since the 1990s (Lin, 1997). As a result, specialised households are dependent on a well functioning marketing system, including the clarity of market signals and the ironing out of distortions (see Chapter 8).

Chinese policy makers see household specialisation as a major way forward for Chinese agriculture, in many regions and agricultural industries. Policy support for specialisation can be facilitative in nature through the development of market systems (to which specialised households are closely linked), the development of local groups and increasing resource mobility.[174] More direct methods of support include: access to preferential loans; the provision of infrastructure (e.g. cattle housing, pens, straw treatment, methane converters); the delivery of extension services (disease control, feeding, breeding); and organisation of the households into supply agreements and contracts with processors.[175]

9.2.2 Budget analysis of specialised cattle fattening households

The economics of specialised households systems are different to those of unspecialised households in a number of ways. Specialised households require a concerted level of household labour, management, land or other household resources dedicated a particular activity, and thereby forego other household activities. The specialisation often means sourcing inputs (such as feed) from outside the household system and generating larger surpluses to sell off-farm. Thus, specialised households are more closely integrated with markets than are unspecialised households. As a result, all items for the representative household budget are assumed to be based on market valuations. In addition, significant household labour is inputted into the specialised activity and this labour can often access alternative off-farm

174 Resource mobility applies to: labour (to allow unspecialised households to reduce agricultural activities which frees up their resources to be used by specialised households); land (to allow unspecialised households to lease more land to specialised households); and capital (such as the development of rural credit markets for specialised households to finance operations).

175 For example, a fattening household visited in Tiandong Village in Shandong was provided with collective land and pens by the village and county government on the condition that it was used for cattle production. Government also supported the households' application for a loan of Rmb10,000 for the purchase of cattle at an interest rate of 0.96% to 1.12% per month.

work. Thus, the "bottom line" for specialised households is more likely to include opportunity costs of labour, and household decisions are likely to be based on "returns to management" (rather than "returns to labour and management").

Longworth *et al.* (2001, Chapter 6) provide a detailed discussion of the economics and budget analysis of a specialised fattening household under regimes common in 1998.[176] The household budget was run again using parameters from fieldwork in western Shandong Province in 2006[177], as shown in Table 9.2. Net profits ("returns to management") for the representative household were Rmb121 per head fed, or Rmb27 per head per month.[178]

Table 9.2: Summary of revenues, costs and net returns for specialised fattening households in agricultural areas, 2006

	Cattle	Other livestock	Cropping	Total on-farm
Total revenues (Rmb)	**97,734**	1,115	3,147	101,996
Own feed			5,840	5,840
Livestock and crop sales revenue	99,648	1,939	11,120	112,707
Total costs (Rmb)	**84,894**	1,995	5,147	88,918
Feed	22,686	1,678		21,843
Feeder cattle	57,600			57,600
Returns to labour, management & capital	**14,754**	–56	5,973	23,789
Returns to management	**4,342**	–502	–106	6,852
Returns per head turned off	**121**			

176 The specialised household had 12 feeders in stock, turned off 36 head per year on a 120 day feeding regime with a daily weight gain of 0.935 kilograms. The household bought in cattle at 300 kilograms at sold them at 412 kilograms.

177 The representative household still turned off 36 head per year, but bought in feeder cattle at a younger age of 12 months at a weight of 200kg (from the unspecialised cow–calf households discussed in Section 9.1), which were fed for 135 days and sold out at 312kg at 500 days old (to slaughter units discussed above or feedlots discussed below), making a weight gain of 112 kilograms, or 0.83 kilograms per day on average. At a dressing percentage of about 50%, the carcass weight is very close to the average in Shandong (officially 148 kilograms in 2005) suggesting that this is the type of beast typically turned off in Shandong. Feeder cattle prices increased (from Rmb6 in 1998 to Rmb8 in 2005), as did finished cattle prices (from Rmb7 to Rmb8.7). Labour prices also increased from Rmb3,125 per year in 1998 to Rmb5,000 in 2006.

178 This equates to "returns to management, labour and capital" of Rmb90 per head per month, which is similar to calculations by the Dezhou Animal Husbandry Bureau officials that returns per head per month excluding labour costs are Rmb100.

The profitability of specialised households is subject to a range of sensitivities and risks. Technical gains that increase the efficiency by which this feed is converted into liveweight gain—due to feed regimes, breeds and management—are important, as shown in Longworth *et al.* (2001, Table 6.4).[179] Fieldwork, secondary data and official statistics suggest that technical efficiencies had not increased significantly between 1998 and 2006 in specialised households. However, one area where there was significant change is the age and weight that the cattle enter the specialised fattening households (from the unspecialised cow–calf households), which decreased substantially from around 24 months of age (at 300kg) in 1998 to 12 months of age (at 200kg) in 2006. This is another dimension of specialisation of household activities, driven by the greater technical and scale efficiencies in specialised fattening households compared to unspecialised households.

Another sensitivity that can be examined is feed prices, which fluctuate significantly over both the short term[180] and the long term as illustrated in Figure 5.7[181]. The fluctuations have a discernable but not major impact on the budgets of specialised fattening households. This is because feed costs constitute 25% percent of all costs in the specialised household and feed regimes are less grain intensive compared to cattle feedlots or pig and poultry operations in China. A detailed sensitivity analysis of feed prices is reported in Longworth *et al.* (2001, Table 6.3).

The most important determinant of returns and source of risk for the specialised households is the price relativity between feeder cattle purchased in and finished cattle sold out, as summarised in Table 9.3. Feeder cattle costs (which constitute 67% of all costs) increased substantially between 1998 and

---

[179] Without changes in the feed regime, if the feed conversion ratio changes from 4.5 to 4 (due to breed improvements, disease control or changes in the age at which feeders are weaned and begin intensive feeding), the average daily weight gain increases from 830 to 933 grams, and net returns increase to Rmb242 per head turned off. This result broadly conforms with calculations of the Dezhou Animal Husbandry Bureau that by feeding 20 days earlier, feeding F2 cross-bred cattle and increasing daily weight gains by 360 grams, profits (not including labour costs) can increase by Rmb340 per head turned off.

[180] Feed prices are reported by specialised household to change by as much as Rmb0.1 or 10% over a three month period.

[181] There are price differences between provinces for corn, the main feed type for fattening households. In September 2005, corn prices were Rmb1.13/kg in Jilin, Rmb1.38 in Anhui and Rmb1.27 in Shandong. These prices are almost identical to average corn prices in 1998 used in the original model. However, there were large fluctuations in corn prices between these two time periods. For example, the average corn price for China as a whole was Rmb0.93 in January 2000 but increased to 1.42 in August 2004.

2006 (by Rmb2/kg liveweight or more than 20%), slightly faster than finished cattle prices, thus applying a price squeeze to the fattening households.[182] Feeder cattle and finished (or slaughter) cattle can be regarded as independent products, with different attributes, markets and buyers and sellers (not unlike for example the different markets for yearling steers and "Japanese bullocks" in Australia). However, markets for generic types of feeder and slaughter cattle can also be related through derived demand and can move together, though with price differentials emerging over the time that the animal if being fattened. Short term movements can also be substantial with reports of Rmb0.2/kg liveweight price shifts over a three month period.

Table 9.3: Impact of input–output price relativities on net returns to representative specialised cattle fattening household—Rmb/head, 2006

	Outputs	312kgLW, 135 days on feed	312kgLW, 135 days on feed	416kgLW, 260 days on feed	500kgLW, 362 days on feed
Inputs	Price unit: Rmb/kg LW	Rmb8	Rmb8.7	Rmb9	Rmb9.3
200kg LW 12m.o feeder (from unspecialised cow–calf household)	Rmb7.7	–38	181	577	824
	Rmb8	–98	121*	517	764
	Rmb8.4	–178	41	437	684
	Rmb8.7	–238	–19	377	624

All returns are expressed in Rmb net return per head
* Price alignments and technical parameters used in representative budget in Table 9.2.

Household feeding is unviable if the unit price of purchased feeder cattle is the same as or greater than the unit price of finished cattle sold out. That is, fattening households must add "value", not just liveweight, to the cattle. Households therefore try to minimise feeder cattle costs (which constitute about two-thirds of their total costs)[183] and to maximise finished cattle revenues (which constitutes virtually all of their revenues). For this reason, the more entrepreneurial households skilled at trading and speculative

---

[182] Indeed, there are cases where per unit feeder cattle prices are actually higher than finished cattle prices. This was observed on fieldwork in Shandong and also appears in some time periods on the market report information for the Tongliao live cattle market.

[183] Cheap feeder cattle are for example sold from the eastern grasslands of Inner Mongolia for about Rmb7.7/kg liveweight, a price which is reinforced in quote prices at the Tongliao live cattle market.

(rather fixed regime) feeding tend to be most successful in the business. Another option is for the specialised household to enter into fixed supply arrangements with abattoirs that offer price premiums for cattle produced to specification (weight, age and dressing percentage). There are substantially higher costs associated with meeting these requirements (especially in feeder cattle and feed costs) but the discussion in Section 8.3 suggested that these costs may be offset by the price premiums *if* contract terms are adhered to and *if* abattoirs do not downgrade product to reduce input costs.

The channels through which specialised cattle fattening households sell cattle are depicted in Figure 9.3. Sensitivity analysis suggests that small mechanised abattoirs (see Table 7.3) have incentives to buy finished slaughter cattle from specialised fattening households if those small abattoirs can access mid value beef markets. Similarly, modern abattoirs (see Table 7.6) will buy 500kg cattle if they can access high value beef markets. It is also relevant to note that the specialised fattening households have a competitive advantage over feedlots in cattle fattening, except possibly for very heavy cattle (see Table 9.5). Feedlots will also buy 312kg feeder cattle from specialised households for further fattening at Rmb8.7 or Rmb9, but only if these are good types of cattle that gain weight efficiently over long term (260 day) feeding regimes.

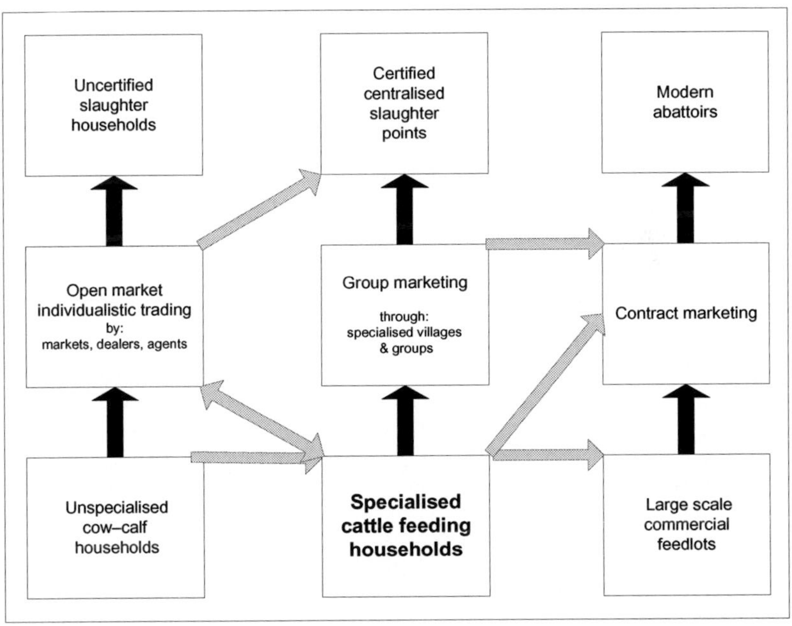

Figure 9.3: Synthesised supply chain for specialised cattle fattening households

The relative viability of specialised household cattle production (Table 9.2) under a range of scenarios (Table 9.3) and the flexibility of production regimes and sales channels available to specialised households (Figure 9.3), has led to the increase in the relative importance of the specialised household production sector. Longworth *et al.* (2001, Table 4.2) shows the increase in specialisation in most surveyed areas in the 1990s. Scale data for 2003 to 2005 presented in Table 5.2 also shows that the relative importance of specialised fattening households that turn off 10 to 49 head and 50–99 head has increased. This contrasts with the contraction of both the unspecialised cow–calf production sector and the feedlot sector in most provinces.

9.3 Feedlots

"Scale" production (*guimo shengchan*) represents the next step up the continuum of scale, technical efficiency and intensity of systems in the production sector. Another point of differentiation between specialised households and "scale" commercial producers is that the specialised household operations are usually owned and managed by the same person, while the owners of "scale" companies usually employ operational managers. The incidence of scale production units in Chinese agriculture is low most fundamentally due to constraints in land size. Most cases of scale production in Chinese agriculture occur on State, collective or demonstration farms. However, scale production is common in livestock industries that are labour intensive and semi-industrial activities. There are large numbers of very large scale piggeries, broiler, layer, and hatchery plants in China.

Cattle feedlots are common throughout China, but are insignificant compared to household producers, as shown in scale data in Table 5.2. There are only 200 large feedlots in China with a turnoff of more than 1,000 head (so on a 120 day feed regime could have a capacity as low as 250 head); roughly 1,000 in the next category down (500–999 head); and 10,000 in the next category (100–499 head). Together these "feedlots" turn off no more than 7% of China's cattle. Table 5.2 also shows that the numbers of large feedlots have reduced for China as a whole, while smaller feedlots—which can also be classed as large specialised cattle fattening households—have increased.[184]

[184] Numbers of feedlots that turn off more than 1,000 head reduced significant between 2003 and 2005 in all areas, especially in the Northeast (Jilin and Liaoning), although Shandong feedlot numbers were stable. The number of feedlots that turned off 499 to 1,000 head per year changed little between 2003 and 2005 for China as a whole or for the northeastern provinces. There has been a substantial expansion throughout China in the numbers of smaller feedlots (which can also be large specialised fattening households) with a 100–499 head turnoff.

9.3.1 Issues in "scale" livestock production

The small size of the feedlot sector is not due to lack of policy support. Scale production is encouraged, especially at local levels. Feedlots are usually classified as "dragon head enterprises" and thereby eligible for a range of concessions. Preferential policies that reduce capital costs (low interest or interest free loans) and reduced taxes and fees have a significant impact on feedlot viability. Land allocated for large scale intensive livestock systems is, under the Livestock Law, treated as agricultural rather than industrial land for tax and zoning purposes. Beef programs especially in the late 1990s and early 2000s—such "Straw for Beef", the "Favoured Area" program and the World Bank Beef Project—included the development of feedlots. Despite—or perhaps because of—this support, the underlying economics of lot feeding has contributed to the relative decline of the sector into the 2000s.

9.3.2 Budget analysis of cattle feedlots

Within the feedlot sector there are different types of feedlots—four or which were identified and surveyed in Longworth *et al.* (2001, Chapter 7). For more details on the feedlots surveys and budgets see Box 5. The parameters and results for the "medium sized feedlot" category were updated in 2006 as summarised in Table 9.4 below. The results confirm previous conclusions that margins in the sector are low given the small design capacity and under utilisation in the sector compared to feedlots in Western countries.

Box 5. Background to the feedlot budget analysis

Longworth *et al.* (2001, Chapter 7) provides a detailed examination of the economics of feedlot cattle production in China in the late 1990s. A total of 30 Chinese cattle feedlots were formally surveyed in 1998, of which 21 were considered suitable for further analysis. Median values of survey results were used to determine both production practices and feedlot profitability. A feedlot budget was developed to compare feedlot costs (feeder cattle, feed, health, transport, labour and various capital costs) with revenues (from the sale of finished cattle and manure) and to estimate net returns per head. This budget analysis was reconciled with survey data and interviews on feedlot production regimes (including feed regimes, time on feed, and weight gains).

Feedlots surveyed appear to provide a good representation of the sector in regions visited. Feedlots surveyed were small and medium sized (design capacity of less than 900 head), rather than large sized which were scarce in fieldwork regions. All the feedlots surveyed are classified as short-term feedlots (feeding for 90 to 120 days), because very few long-term feedlots (that feed for 240 days) were evident. Surveyed feedlots are all located in Shandong, northern Hebei and Liaoning provinces. Five of the eight medium sized feedlots were integrated with abattoir facilities. All the feedlots surveyed were "regional" fattening units, supplying "regional" abattoirs. None were part of a large, modern, purpose built, integrated feedlot-abattoir facility. Based on the characteristics of the feedlots, they were classified in four different categories: Medium sized commercial feedlots; Small sized commercial feedlots; Households in collective feedlots; and Household feedlots. The budget analysis presented in Section 9.3.2 is for the medium sized commercial feedlot.

This survey information from 1998 was updated using information collected on fieldwork in 2006. This fieldwork included interviews with feedlots and various industry experts and officials. The new parameters were inputted into the previous feedlot budget spreadsheet to generate the budget and sensitivity results presented in this chapter.

Table 9.4: Profit & loss statement for representative medium sized feedlot, 2006

	Rmb/head	Percentage
Revenues	3,802	100%
Revenue from finished cattle sales	3,742	98
Manure sales	60	2
Costs	3,835	100%
Feeder cattle	2,714	71
Total feed costs	843	22
Transport	80	2
Depreciation costs	32	1
Interest payments	46	1
Taxes and utility costs	25	1
Wages	74	2
Health and inspection costs	20	1
Net revenue/profit	–33	

Parameters: Design capacity 100 head; 100% utilised; 434 head turn off per year, feeder cattle enter 312kgLW (at Rmb8.7/kgLW), fed for 108 days at an average daily weight gain of 1.25kg, finished slaughter cattle sold out at 416 kilograms (at Rmb9/kgLW). Feed regimes specified in Longworth *et al.* (2001, Table 7.2).

Like specialised households, the profitability of feedlots is sensitive to technical efficiencies and feeder/finished cattle price alignments. However, compared to specialised fattening households, feedlots have higher scale and technical efficiencies and lower transaction costs. They have higher input costs such as feed (because of the more concentrated feed regimes), labour (which must be hired in), as well capital costs and overheads, the latter of which is related to capacity utilisation. As a result of these higher cost structures, feedlots need to access higher value markets. This can be achieved through access to export quotas to Hong Kong or on the mainland market through their capacity to ensure quality and continuity of supply. In the absence of these premiums, feedlots can not to compete with specialised households that operate on lower cost structures. As a result, abattoirs source most of their cattle from specialised fattening households. An overview of the supply chain for most feedlots in China is illustrated in Figure 9.4.

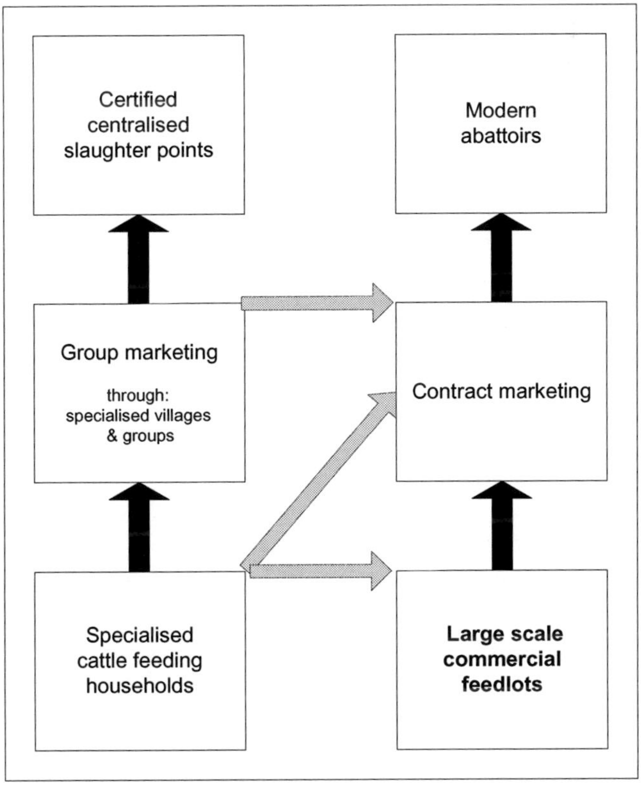

Figure 9.4: Synthesised supply chain for feedlots

While feedlots use more intensive feed regimes than specialised fattening households, feedlots also have higher cost structures for other items (e.g. capital costs, overheads and depreciation). As a result, feed costs account for a smaller proportion of overall costs (22%) than is the case of specialised households (25%). Feed costs are also likely to constitute a lower proportion of overall costs for cattle feedlots than in the case of large sale pig, broiler and layer operations. Nevertheless, feedlot returns are sensitive to changes in feed prices as overviewed in Longworth *et al.* (2001, Table 7.4). In the years in which the feedlots were budgeted (1996 and 2006) feed prices were similar (see Figure 5.7). Weight gain efficiencies are another important determinant of feedlot profitability.

Returns are most sensitive to the differentials between feeder cattle prices and finished cattle prices as shown in Table 9.5. The feeder cattle inputs bought into to the representative feedlot (500 days old, 312kg) are sourced from the representative specialised fattening household (Table 9.2) at various prices. Well managed feedlots avoid purchasing poorly bred, overly light cattle (for their age), or overly fat cattle that have a low capacity to gain weight in the feedlots. Feedlots will however seek (at lowest cost) to source "good value" cattle that have the potential to gain weight efficiently. Furthermore, feedlots have a strong preference for homogeneous lots ("even lines") of this type of cattle in order to reduce transport and marketing costs and to increase management efficiencies in the feedlot (so that a whole line of cattle can be fed similar feed rations to maximise weight gains).

The outputs row of Table 9.5 contains a range of information or choices for the feedlot. Most fundamentally, the feedlot must choose the feed regime that it will adopt, especially the time on feed. While "good type" (for example F2 or higher) improved cattle gain weight and therefore increase revenues over time, diminishing marginal weight gains set in at heavier weights which must be offset by significant (per unit) price premiums. Under the assumptions used in Table 9.5 these premiums appear sufficient to provide incentives to feedlots to produce heavy cattle. However, it is important to note that it can be difficult to viably sell these heavy cattle. Table 7.6 shows that modern abattoirs can only buy heavy cattle (500–600kg) at these prices (Rmb9.3–9.6/kg) if they area able to access high value beef markets. As argued in Section 7.3, this can be very difficult for most modern abattoirs, which severely limits the market that feedlots have for heavy cattle. If the feedlot does not have a viable sales outlet for this finished beast then it must adopt a shorter term feed regime (90 to 150 days) which may be profitable if the per unit input price of the feeder cattle is lower than the per unit output price of the finished cattle (that is, the "value" of the animal is increased).

Table 7.3 indicates that small mechanised abattoirs appear to present a ready market for this type of animal. However, the feedlot is less profitable—and therefore at a competitive disadvantage to—the representative specialised household shown in Table 9.3 in the supply of these lighter types of cattle. This analysis suggests that unless there is substantial growth in the higher value beef market segment, feedlots will not take over from smaller household producers, which is a commonly voiced concern in the literature. Furthermore, policy measures to support feedlots will not be cost-effective or sustainable.

Table 9.5: Impact of input–output price relativities on net returns to representative medium sized feedlot, 2006

	Outputs	416kgLW, 83 days on feed	416kgLW, 83 days on feed	500kgLW, 150 days on feed	600kgLW, 230 days on feed
Inputs	Price unit: Rmb/kg LW	Rmb8.7	Rmb9	Rmb9.3	Rmb9.6
312kgLW, 500 days old	Rmb8	61	185	393	625
	Rmb8.7	–158	–33*	175	406
	Rmb9	–251	–127	81	313
	Rmb9.3	–345	–220	–13	219

* Price alignments and technical parameters used in representative budget in Table 9.4.

Further insights into the underlying economics and the management skills of matching feeder cattle inputs, finished cattle outputs and production regimes are provided in Table 9.6 which is a summary of interview information with feedlot managers in northwestern Shandong (Dezhou) in 2006. The feedlot had switched from buying feeder cattle from nearby northeastern Shandong (Laizhou) to the much more distant area of Liaoning (Beining). There are large numbers of livestock agents in Beining that provide purchase, trucking, inspection and fee paying services. Even though these set of purchase costs are Rmb70 per head higher than buying locally, the costs are more than offset by the returns. Beining was said to be a relatively poor area so feeder cattle prices are low and a breeding program has been underway in the region for nearly two decades so young cattle responded well to feeding. The managers estimated that while purchasing feeder cattle from nearby Laizhou brings about a loss, the cattle from Beining bought profits. Incentives such as these have caused a dramatic expansion in inter-regional cattle trading, as mentioned in Chapter 8.

Table 9.6: Feeder and finished cattle price relativities and impact on profitability for a feedlot in Western Shandong

Year	Feeder cattle				Finished cattle	Feedlots' perception of profitability
	Feeder cattle source	Transport and agent fees (Rmb/head)	Breed of cattle/feed conversion	Purchase price (Rmb/ kg LW)	Sale price (Rmb/kg LW)	
2003	Shandong	180	3 (Luxi-Limousin)	8	7	Loss
2004	Shandong	180	3 (Luxi-Limousin)	9–10	8.4	Loss
2005	Shandong, Liaoning	—	—	9	9	Broke even
2006	Liaoning	250	3.3 (Simmental-Charolais)	8	8.6–8.7	Profits

9.4 Concluding comments

An analysis of the structures, economics and policies of the cattle production sector enables findings that may be of interest to Chinese policy makers and to researchers on small-holder agriculture. Small scale unspecialised households are incrementally exiting semi-subsistence agriculture as they allocate labour to other more lucrative activities, mainly off-farm. While abattoirs and livestock and trade officials are concerned about the consequences of this for agricultural output—resulting in subsidies of sow production for example—measures that effectively maintain small scale unspecialised agriculture are misplaced if the objective is to increase rural incomes over the longer term. At the same time, preferential policies toward large modern feedlots will be counter-productive in most cases because they are unable to compete with smaller specialised fattening operations in all but the production of very heavy cattle for which there is a limited market because of the small size of the premium beef market. Specialised household production structures provide a viable, flexible and inclusive model of development to service a range of markets, but especially mid value markets and especially if the households participate in local groups. As a result, rather than directing policy attention toward unspecialised households or large scale feedlots, it is argued that policies should be directed at increasing the mobility of resources (land, labour, capital) to facilitate households specialisation.

Chapter Ten

Recommendations and Policy Implications

The modernisation of Chinese agriculture has proceeded at pace but has certainly not proceeded smoothly, as illustrated through the detailed analysis of the beef industry in this study. The analysis provides a strong information base from which to consider how China's agricultural modernisation strategy could be refined or indeed recast. Policy recommendations in this chapter are structured around a sequential discussion of low, mid and high value agrifood chains, each of which cut through the sectoral (beef market, cattle processing, marketing and production) chapters of the book.

It is argued that policy makers, industry, researchers and development agencies are justified in their interest in the modernisation of Chinese agriculture. However, there is a tendency to over-emphasise the speed by which modernisation does and should take place, especially if the modernisation strategy aims to skip development phases to build "high value agriculture" structures in a way that is contrary to market forces and the interests of industry participants in rural areas. It is argued below that a more sustainable and inclusive approach to modernisation is required, based around incrementally upgrading existing structures and the development of mid value markets and agrifood structures. Amongst the justifications for proposing this "mid value agriculture" modernisation strategy is that it is better aligned with both market forces and the interests of industry participants in rural areas than is a modernisation strategy based on over-emphasising high value markets or, opposingly, maintaining the continued and almost complete dominance of low value agrifood chains in Chinese agriculture.

It is important to note however that this proposed modernisation strategy is not an either/or proposition. Low, mid and high value segments coexist in China's increasingly segmented agricultural landscape, in different proportions in different regions, while there are also strong inter-relationships between the various value chains and their participants. Thus, the modernisation strategy proposed below entails more of a shift in emphasis, or an "evolution" toward "mid value agriculture" rather than a wholesale transformation, or a "revolution" toward "high value agriculture". In addition, the impacts of policy measures will not be confined to discrete value chains but will impact on all value chains to some extent. The policy recommendations made below are not discriminatory toward or against particular value chains. To the contrary, the chapter argues the need for further development of "market facilitating" and "public good" policies such those toward food safety, standards and the development of markets and argues the need to reduce or abolish anti-competitive policies and preferential treatment. These market facilitating and public good policies are closely aligned to a "mid value agriculture" development strategy.

10.1 Low value agrifood chains

Any discussion on agricultural and rural development in China must be based on a close understanding of low value agrifood chains. Low value agrifood chains provide the foundation on which Chinese agriculture will modernise, are the source of low cost food for consumers and are the source of livelihoods for an enormous number of rural people. The modernisation of Chinese agriculture will impact significantly on these low value agrifood chains, their participants and consumers. Despite the importance of low value agrifood chains, they tend to be undervalued and overlooked in China's modernisation drive. Many policy makers in China pejoratively describe low value chains as small-scale, "chaotic", "fragmented", "unstandardised" and "backward". While these terms may be an accurate depiction of low value agrifood chains in China, it is also a mistake to attempt to replace them with large and highly centralised structures that display polar opposite characteristics. Rather, an incremental development of the low value agrifood chains is required.

Markets

With China's economy forging ahead, there is a growing demand for animal protein and other food products with positive income elasticities such as beef. While the consumption of beef as a broad product category has increased significantly in recent years, the vast majority of beef consumed is still low value with few quality characteristics and variable hygiene levels. This generic beef fills the expectations of the vast majority of Chinese consumers, especially as high heat cooking and the heavy use of flavourings negate many hygiene and quality issues. In this regard, low value beef chains are effective in delivering an affordable, lean source of protein food to consumers.

Against this sanguine perspective of low value agrifood chains, however, there are pressures to improve basic food safety standards of agricultural products in China which will require the further regulation and restructuring of agrifood chains at retail, slaughter, marketing and production stages. A rapidly growing population of more affluent urban consumers are prepared to pay for basic food safety and quality standards, especially where lighter cooking methods or presentation is required (for hot pot for example). An increasing number of cities are also regulating and restructuring low value agrifood chains to improve food safety standards. Increasing demand for basic food safety and quality attributes in food are slowly and incrementally increasing the relative importance of mid value beef markets in the overall food market, which in turn impacts on the agrifood chains that deliver the products.

Agribusiness

One of the outcomes of China's reforms of the 1980s and 1990s was the rise and dominance of highly decentralised and deregulated agribusiness and marketing sectors where product is sold, processed and traded in small scale, individualistic structures, largely devoid of effective regulation, inspection or registration, let alone the application of quality standards.

These rudimentary low value agrifood chains play a valuable social role. Participants of low value agrifood chains are flexible, competitive and work in adverse conditions to get food to market fresh every day of the year at low cost. In addition to benefits to consumer welfare, low value agrifood chains are highly accessible and provide channels of mobility for large numbers of rural households to move off-farm and into the light industry and services sectors. In the cattle, sheep and goat meat industries, a large proportion of these rural people belong to ethnic minorities. Budget analysis in this study indicates that slaughter households and the traders of cattle, beef, offal and hides can become "moderately well off" (*xiaokang*) through these off-farm activities. Returns invested by the households in education, health, business investment and further mobilisation up the supply chains create multiplier effects to the process of rural development.

While the low value agribusiness sector is both inclusive and labour intensive, it has a poor record in the delivery of safe or quality assured product. As a result, China has in recent years endeavoured to better regulate and/or restructure the sector. With regard to the regulation, compared to the rudimentary policy settings of the 1990s (see Waldron et *al.*, 2003a,

Chapters 11 and 13), China has in recent years issued an enormous number of policies, laws, regulations and food safety and quality standards for food. Of most relevance to low value markets are basic national standards, rather than more sophisticated certification schemes. Measures to strengthen national regulations have been complemented by measures to strengthen institutional capacity to enforce the regulations. Especially important institutional reforms in this regard include the elevation of "macro control" departments such as the Industry and Commerce Administration Bureau, the State Food and Drug Administration, and the Ministry of Environmental Protection and the Veterinary Bureau (see Waldron *et al.*, 2006).

While China appears to have taken significant steps to better regulate low value agrifood chains, effective enforcement of the regulations is a very different matter. For example, more than a decade after regulations to ban household slaughtering were issued, they have simply not been implemented in the beef industry in most areas of China, while disease and hygiene inspection in markets is still largely a rubber-stamping process. As a result food (and drug) safety problems continue to emerge regularly in China, most recently in melamine contamination of milk by households and milking stations. It is also widely lamented in China that there is a glaring institutional void in the area of agricultural marketing and that the vast majority of trade occurs in the absence of consumer or producer protection, contract enforcement, or objective grading and testing.

However even with increased regulatory powers and enforcement capacity, the sheer size and the decentralised (or fragmented) nature of low value agrifood structures precludes effective and comprehensive regulation. It is either logistically impossible or prohibitively expensive for local authorities to know about, inspect, or pursue complaints about the activities of an enormous number of traders and processors in their constituency. Even if local authorities can track the activities of processors and traders, it can be very difficult for local government to recover costs. Local authorities can be unwilling to levy—and slaughter households and traders can be unwilling to pay—fees and taxes, especially if the slaughter households and traders are ethnic minorities. Central government regulations to cut agricultural taxes, recent calls to cut inspection fees related to food safety, and inadequate transfers from central government to compensate for the loss of these revenues, further reduce the incentives for local government to effectively regulate fragmented agribusiness structures.

Efforts to improve food safety and market conduct must therefore necessarily extend beyond regulation to include restructuring of the industry itself. Closing or merging small agribusiness structures enables centralisation and concentration into larger structures that authorities can monitor and inspect more cost-effectively. Small agribusiness structures must also upgrade facilities to meet more rigorous registration and certification criteria, including those for capital investment, infrastructure, and hygiene and effluent managements systems. There is also a degree of self-regulation in abattoirs in China as inspectors are generally employed by, and wages are paid by, abattoirs. Restructuring of the beef industry has the potential to improve food safety (as is the case for pig and poultry slaughter), to improve effluent management (as is the case for leather and paper industries) or to realise efficiencies in aggregation (as is the case for pig slaughter and milking stations). Indeed, these more centralised structures are already evident in the beef industry in more developed parts of China or where there are large ethnic minority populations that include beef as a staple part of their diets.

Given the movement toward increased regulation and restructuring, major questions arise about the degree to which low value agribusiness structures should be centralised. In particular, consideration has to be given as to how to minimise costs on the displaced slaughter households and traders and how to minimise costs for price-sensitive consumers and thereby avoid pricing beef out of the market. Again, these choices will be made by business and government actors at a local (especially city and county) level based on their own incentives and in the context of levels of economic development in that local area. However, the decisions of local business and government agents are also impacted by policy settings at higher (central and provincial) levels.

It is argued in this study that a viable and inclusive model by which to regulate and restructure the agribusiness structures of the beef industry is through the introduction of a significant number (up to 10) smaller scale "unmechanised centralised certified slaughter points" in any given city. Compared to slaughter households, unmechanised centralised certified slaughter points are less costly to regulate and, compared to large mechanised abattoirs, represent a lower cost and more viable structure from which to service low value markets, or even to service mid value markets for some cuts. The unmechanised centralised certified slaughter points are also more labour intensive than large mechanised abattoirs. Indeed, they are often service slaughter operations or simply collective facilities for the use of individual slaughter households. This model is prevalent in large cities such as Beijing and acts as an effective model for other cities in China.

Despite the appropriateness of this relatively decentralised slaughter model, a more centralised model is becoming predominant in the cattle slaughter and beef trading sectors in China. Cities such as Changchun, Urumqi, Altai, Ji'nan and an uncountable number of counties throughout China have or are in the process of granting preferential licences and market access to relatively large scale mechanised structures, which are either refurbished ex-State owned abattoirs or new large modern abattoirs. The main reasons why local government pursue this model is because they are easier to inspect and tax, because they make use of unutilised or under-utilised facilities (especially ex-State owned abattoirs and cold storage facilities), and because government can gain from patrimonial relationships with big business.

The main problem with this overly centralised slaughter model is that it establishes local monopolies or oligopolies in the slaughter sector of the city or county, which can extend through to the control of beef distribution and storage in food markets in the city or county. The companies are supported not only by preferential market access but also by preferential terms for the privatisation of ex-State owned abattoirs and subsidies to stabilise meat prices in areas with large Muslim or Mongolian populations. The local monopoly or oligopoly status means that the livelihoods of all industry participants in the area—from cattle producers to slaughter workers to beef traders and stallholders—are subject to the competitiveness, underlying viability, management and conduct of the operations. Placing business operations (especially abattoirs) in this position poses high risks to the full range of industry participants.

Production

Even with modest margins in the low value agribusiness sector, the low prices in generic beef markets means that returns to primary producers are also low. This is especially the case given the small sizes and low technical levels of the household producers that sell into low value agrifood chains. Low levels of development in many parts of rural China (including low opportunity costs of labour) and only partial integration into markets (for semi-subsistence households), partially mask these low returns. For example, many semi-subsistent households that hold cattle undervalue or do not value at all major cost items in primary production such as labour and feed.

In the 1990s, interventionist policies toward the production sector impacted strongly on these poor and semi-subsistent households. Interventionist measures included the organisation and "encouragement" of agricultural households by local government (to produce particular types of products), inducements to participate in programs (such as the Straw for Beef program), the use of the State extension system as an instrument of policy (breeding, feeding, veterinary) and in some cases subsidies to stabilise prices (for grain and pork production, for example). As outlined in Waldron *et al.* (2003a, Chapter 9), interventionist policy toward the production sector was designed to build a "production base" on which the industry would develop. This objective was successfully achieved and indeed it could be argued that it was "over achieved" causing an over supply of cattle in the late 1990s.

Compared to the 1990s, these types of interventionist policies now have become increasingly blunt instruments by which to influence household production decisions. This is especially the case in the more developed agricultural areas of China. Access to more lucrative off-farm work and the application of market based opportunity costs to agricultural production means that households are increasingly ignoring the subsidies and "official encouragement" offered by local government. At the same time, interventionist policies applied to primary production have switched from beef cattle to other industries such as dairy and sheep meat, especially in more remote parts of China.

These economic, policy and social forces have bought about a relative decline—and in many cases an absolute decline—in the small scale unspecialised production sector in China for a large range of agricultural products. This is a source of considerable concern in China. Food and fibre processing industries are concerned about the impact on the supply and prices of agricultural inputs for processing, while policy makers are concerned about the impact on food prices and trade balances. The response in 2007 and 2008 was to provide subsidies for the production of strategic and staple commodities like grain and pigs. In the case of the latter, households breeding operations—most of which are small households—are provided with cash subsidies for the number of productive sows they raise. Representatives of the Chinese beef industry (abattoirs, the industry association and researchers) lobbied central government in 2007 and 2008 to extend the same type of subsidies to cow–calf producers. This specific proposal was rejected by President Hu Jintao and the Minister for Agriculture, because the beef industry is no longer a priority or strategic industry for central government. However, some local governments—even in the Central Plains—still proactively try to build or maintain cattle "production bases" for the perceived benefit of producers, consumers or abattoirs.

Measures at industry or local levels to arrest the decline in the small scale unspecialised production sector are misplaced if the objective is to improve rural livelihoods. Intervention in this unspecialised household production can delay the movement out of low value unspecialised agriculture, can lead households into bad investments, or can create conditions of over-supply, all of which prolong a poor life for the households. Small scale unspecialised households in more developed areas of China should be provided with an environment in which they can, over time, either exit agriculture altogether or to "step up" into becoming specialised household producers.

This approach will be less relevant in remote and undeveloped parts of China where small scale, semi-subsistent, unspecialised households continue to play an important role in food security and risk diversification. Many households in undeveloped parts of China still have incentives to raise cows due to low opportunity costs of labour and the continued use of cattle for cultivation. In these cases, calves contribute to household incomes, provided that the calf is weaned and sold at a young age to specialised fattening households.

As livestock activities given up by relatively wealthy rural areas and households are taken up by relatively poor rural areas and households, cattle production—and indeed livestock production more generally—may "move west" in a so-called "flying duck" formation. There is some evidence (in scale of production data, various surveys and information from large live cattle markets, for example) that an increasing proportion of cows and calves are being produced in the more inland areas of China, and that a greater proportion of calves and feeder cattle are transported to more developed areas for finishing and slaughtering closer to large end markets. It is also possible that while low value segments of the industry will move inland, that the higher value segments of the industry will consolidate in more developed areas closer to higher value end markets. Spatial change will be determined not only by labour costs, but also feed costs, cold storage and transport infrastructure and policy restrictions on the movement of live animals and meat.

Due to all these factors, spatial change is likely to be a process of regional specialisation of particular sectors and segments of the industry rather than shift in the industry as a whole. Provinces such as Henan and Liaoning are, however, likely to remain major producers across a range of sectors and segments into the future, especially because in addition to containing pockets of poor areas they also have easy access to major urban markets.

10.2 Mid value agrifood chains

The discussion above assesses some of the industry restructuring and policy options in low value agrifood chains needed to safely "feed the masses"—both producers and consumers—into the future. However, the emergence of a sizeable new middle class in China has given rise to the emergence of mid value markets. China has little experience in organising industry to meet the demands of this middle class and mid value market. The response to date has been one of forcing the pace of change in the industry through rapid modernisation and intervention. It is suggested below that a more viable and inclusive response is to incrementally develop "mid value" agrifood chains, especially through the development of facilitative policy and institutional settings.

Markets

One of the major developments in the Chinese beef industry—and other agricultural industries—in a period of little more than a decade has been the development of a significant mid value food market. The rising urban middle class have increasingly diverse diets that include beef and have evolving styles of eating and cooking. They also eat out more including in ethnic or foreign type establishments where beef can hold an important place in the menu.

Filling this demand requires beef with various types of food safety and quality attributes. A critical aspect of building this demand into the future is assuring consistency of supply to reduce purchase risks for customers and consumers. This demanding task may actually be more feasible in mid value markets than in high value markets where large companies have an incentive to differentiate their products through brand names and advertising. In mid value markets, however, smaller and more anonymous companies may have higher incentives to use industry-wide or public standards.

Industry participants that service these mid value markets—especially abattoirs and fattening operations—may benefit from industry activities such as the promotion and assurance of beef as a safe and healthy meat, the widespread adoption of beef quality standards (a beef grading system), beef education campaigns and cooking classes. Indeed, activities such as these are an important aspect of the sustainable development of the industry. However, further research beyond the scope of this study is required to determine the benefits of these activities in relation to the costs from an industry-wide or from an enterprise-specific perspective. Even if benefits outweigh costs for all or any of these activities, major issues of free-riding and coordination amongst industry participants arise. Current industry peak organisations that notionally have jurisdiction over such activities (such as the China Cattle Association and the China Meat Association) do not have the membership, funds or the power to do so. However, given that these activities are vital to the sustainable development of the industry, consideration might be given as to whether a peak industry organisation should be mandated to extract levies for the industry activities, or for government to divert funding spent on subsidies for producers and processors, to be spent on increasing the adoption of food safety and quality standards and other industry development activities.

More specific issues of industry coordination and governance apply in food safety schemes. The many food safety certification schemes in China can be ordered along a continuum from basic national standards that are most relevant to low value markets, to sophisticated Organic Food standards that are most relevant to specific high value domestic and export markets. Of most relevance to mid value markets are the *wugonghai* (Non Public Hazard) Food scheme and the Green Food scheme, both of which are administered by agencies attached to the Ministry of Agriculture, and both of which are widely used in China's food sector. It is reported that consumers are losing confidence in the schemes. This is because certification agencies have attempted to rapidly expand the "market share" of the certification schemes and to attract large numbers of users and revenues for the agencies that are under pressure to develop self-funding schemes. Efforts to rapidly expand the schemes can take place at the expense of rigour of the administration of the schemes. Loss of consumer confidence in the schemes diminishes the willingness of consumers to pay the premiums that are necessary to offset the extra inspection and production costs. Loose management of the certification schemes has implications for rural and regional development because standards such as Green Foods and Organic Foods for ruminant livestock act as a sorely needed source of competitive advantage for more remote pastoral areas of China. Given these issues, the organisations that manage the certification schemes may be better served by maintaining more rigorous and viable—and fewer—certification schemes and scaling back ambitious expansion plans to more manageable levels.

Agribusiness structures

Given the growth of mid value markets in China, important questions arise about the most appropriate agribusiness structures to service the markets.

Again, cattle slaughter is a critical sector in this regard. It was argued above in the discussion on low value agrifood chains that some unmechanised certified slaughter points will be able to sell some of their cuts into mid value markets. This is particularly the case as the more centralised nature of their operations (compared to individualistic slaughter households) allows for the aggregation of particular cuts or types of by-products into larger commercial consignments. This then allows the slaughter units to enter into larger and higher value sales channels for these cuts and raises the average value of outputs.

However, consistently servicing mid value markets with a larger proportion of the carcass requires purpose-built mechanised abattoirs with more developed facilities and systems, including cold storage, packing and processing facilities and well managed slaughter and butchering lines. The question then arises as to the most viable form of abattoir to meet the needs of mid value markets. The large number of interviews and field visits conducted for this study reveal an entrenched bias amongst policy makers at both higher and local levels of government toward the development of large modern abattoirs. The analysis in this study argues against such a "bigger is better" approach and that small mechanised abattoirs provide a more viable and inclusive model to service mid value markets for several reasons.

First, small mechanised abattoirs have lower cost structures than large modern enterprises (especially given low capacity utilisation in large abattoirs) and so can more viably service mid value markets. Second, smaller slaughter units are relatively labour intensive, are usually located in rural areas, and have the capacity to absorb large numbers of slaughter workers from slaughter households when they are effectively banned. Third, smaller and more viable slaughter operations are more likely than larger and less viable modern abattoirs to establish and preserve close, ongoing and localised cattle purchasing channels. The development of trust, communication and mutual benefits between abattoirs and cattle producers about attributes and prices is critical to industry development and the equitable distribution of benefits amongst industry participants.

These small mechanised abattoirs will inevitably be supplied by specialised cattle raising households operating in localised groups with the capacity to assemble sizable and "even" lines of cattle. The functions of these local groups are not just to market cattle outputs, but also to coordinate production inputs, information, capital, disease control, breeding and effluent management. Recent government measures to promote local groups especially through the development of infrastructure for small livestock areas and the passing of the Cooperative Law, are important but not sufficient steps in the development of local groups. The key to the ongoing success of the groups will be in their ability to manage and coordinate their own affairs. More proactive activities for local groups include the slaughter of their own cattle (and therefore the marketing of their own beef) through service slaughtering arrangements or direct ownership or management of local slaughter points.

Contrary to some observations in the literature, successful producers and groups need not necessarily—and in fact do not often—enter into formal supply contracts with specific abattoirs. This is partly due to bad experiences with formal contract systems in the past and a preference to maintain flexibility in sales options and channels. Producers must, therefore, be presented with a choice of marketing channels. In this regard, it is important to upgrade the facilities and services provided by some physical market places so that they remain relevant to higher value trade. Although auction systems have been used for the sale of breeding animals including cattle (and in other commodities such as wool), they remain insignificant in Chinese agriculture. The development of a competitive auction system in particular sites through trials and projects would appear to be worth close consideration from abattoirs, industry bodies and government. It will also be important to increase trust in "over the hooks" purchasing, especially in assuring that slaughter results are accurate and reflected in payments to producers. In the case where disease and hygiene inspectors are appointed and paid for by the State, consideration could be given to expanding the role of these inspectors to the inspection and auditing slaughter/carcass test results, and delivery of the reports back to producer groups. Initiatives to improve marketing systems—such as auctions systems and the reporting of slaughter results—are neglected in China but with continued economic development provide potential to increase pricing accuracy in the industry to the benefit of both abattoirs and producers.

Producers

Specialised households can be seen as a "middle line" form of production structure, lying somewhere in between unspecialised households and large scale corporate feedlots. Specialised household producers will continue to grow in relative importance in agricultural production, including in beef cattle production. However, important questions remain about the type of activities that specialised households will engage in into the future.

Specialised households are most active in cattle fattening. They have higher technical levels and are more efficient in fattening than are unspecialised households, a competitive advantage that will become increasingly apparent as unspecialised households apply market based costs to their cattle production activities. At the same time, specialised households tend to have lower (infrastructure, capital, labour) cost structures and are more flexible than feedlots. The capacity of specialised households to adjust regimes (such as time on feed) to take advantage of price alignments (for various types of cattle and feed prices for example) is especially important. However, this requires that households build expertise in cattle production and cattle markets, which only the more progressive households will be able to achieve. In this regard, training and extension programs that cater more specifically for more specialised producers—rather than the "generic" services of the prevalent State extension system—are important. While the costs of providing these services for local and higher levels of government can be significant, they are likely to be more cost-effective than the subsidisation of large scale abattoirs and feedlots.

It is also important to note that specialised household production has very often been a vehicle by which households "step out" into downstream, off-farm activities like cattle trading, slaughter or beef trading. Mobility paths into the individual slaughter and beef trading sectors will be blocked as the household slaughtering ban is enforced. As discussed above, however, the development of unmechanised slaughter operations provide locations through which householders can still slaughter and market product—if those slaughter operations are collectively owned and managed. There are also prospects for successful specialised households to expand scale of production and to register as companies to form feedlots for example, in which case at least they are doing so from an existing base of knowledge and experience in cattle fattening that can be missing in start-up corporate feedlots. Other opportunities are also available through the development of local groups, which are nearly always comprised of specialised household producers. In this regard, measures in China to develop local groups and the links between primary production, marketing and processing will also impact strongly on household specialisation.

One of the most important and fervently discussed questions in the Chinese beef industry is what type of production structure that will produce China's calves into the future. At present, cow–calf production in agricultural parts of China is almost totally dominated by unspecialised households. However, as noted above, incentives for these households to engage in cow–calf production as an unspecialised activity are diminishing.

Specialised households may be competitive with unspecialised households in the cow–calf sector if they can improve on current low technical levels in areas such as breed and animal selection, nutrition, conception rates, lactation, disease control and early weight gain. Specialised households are more likely to enter the cow–calf sector in a significant way with an increased demand for better quality feeder cattle inputs into the cattle fattening systems. Specialised household are already active in piglet production for higher value pork supply chains. Furthermore, diminishing incentives for unspecialised households to engage in cow–calf production in agricultural systems may shift the competitive advantage of cow–calf production back toward larger and more specialised households in pastoral and semi-pastoral systems in provinces such as Inner Mongolia.

On a more general note, the major means of increasing household incomes in rural China will continue to be to increase opportunities for both off-farm work and for on-farm specialisation. Both of these objectives are served by complementary policy settings, namely to increase the mobility of resources (land, labour and capital). While land rental markets have grown since 2000, many households are still wary of collectives reallocating leases and of government resuming land if it is not used for agriculture. Long term labour movement is still constrained by the household registration (*hukou*) system and the capacity of households to claim social services outside their home towns. Both of these factors restrict the structural adjustment and *zhuanyi* of agricultural households that frees up resources and provides opportunities for specialisation for the households that do stay on the land.

Increased specialisation also increases the exposure of the households to various types of risks. The development of effective insurance systems, information systems and rural capital markets in China are therefore important in facilitating the household specialisation process. The development of information systems to allow rational assessment of market movements and risks are particularly important. Most of the information available to households at present is through the official extension system and is technical in nature. Market information is usually aggregated, including that on output statistics, prices, or companies. Given the difficulties in overhauling the official extension system, some novel alternative solutions have emerged. Many households and associations in rural areas access information through the internet and online market places, some of which are reliable. More importantly in remote areas, two of China's largest telephone companies collect and report disaggregated price information for a range of agricultural products (in order to attract telephone calls and charges from both buyers and sellers). While this telephone service has not yet been linked to price reporting for beef to the extent of reflecting significant grade differentials, the development of such a system would appear to entail important benefits for the industry, especially specialised household producers.

10.3 High value agrifood chains

"Modernisation" in China is generally interpreted as building large scale, high technology and high value structures. Policy makers believe that the modern structures are best suited to serving China into a future of rapid economic growth, with perceived benefits for efficiency, trade, food safety and environmental management, and especially for farmers that can integrate into the systems. In anticipation of these conditions and benefits, China has attempted to fast-track the development of modern agricultural structures. In this regard, there are distinct similarities to the fast-tracking strategies pursued by other East Asian countries through close business-government relations and corporatist structures (see Waldron *et al.* 2003a, Chapter 10).

While there are examples of successful and viable modern structures that service high value beef markets in China, efforts by many local areas to replicate these examples have been spectacularly unsuccessful. The main constraint to the development of the high value agrifood chains on a widespread scale is the small size and the limited growth prospects of high value markets. The absence of a significant high value market has, however, not deterred policy makers and investors from building modern facilities and systems through interventionist and preferential policies, thus creating over-crowded and over-supplied high value agrifood chains. The consequent unviability of the enterprises in the high value agrifood chains creates a series of further problems for downstream participants, including household producers.

Markets

The main factor constraining the development of high value agrifood chains is the small size and limited growth prospects of premium markets. The assumption that premium markets will grow rapidly with economic development in China has not come to fruition for many food and fibre products in China. In the case of beef, wealthy Chinese people do not appreciate steaks or big pieces of meat that are the cornerstone of the premium beef markets in other countries (even if the large pieces are then prepared into thin slices as is the case in Japanese a lot of Japanese beef cuisine). That is, Chinese consumer preferences negate the need for high quality and large portions of beef, or grain-fed and heavy cattle.

Looking into the future, it is also hazardous to assume that high value markets will develop "organically" simply as a function of broader economic development. As discussed above, the development of markets require proactive and coordinated measures in areas such as food safety and quality assurance and promotion and consumer education activities. These activities are unlikely to happen in the premium beef market because large abattoirs seek to differentiate their products rather than coordinate to develop the market. The conduct of large abattoirs—including false advertising, fake products and lack of quality assurance—also deters customers and consumers in the premium market from sourcing beef domestically.

The small size of the premium beef market and the embedded nature of the constraints to its development mean that an inordinate number of enterprises in high value chains are unviable. The viability of high value agrifood chains depends on sufficient price premiums for beef in order to offset the higher cattle and other costs. The premiums have not been sufficient in the case of beef, or many other commodities, for the establishment of viable and sustainable high value agrifood chains. In this regard, the cattle slaughter sector may be different to some industries like chicken slaughter and milk processing and packing where size and technical efficiencies (rather than price premiums) are major determinants of competitive advantage, and which allows these structures to viably service lower value agrifood chains.

Agribusiness

High value agrifood chains—sometimes referred to as agro-industrialised systems or modern supply chains—generate enormous interest from policy makers and researchers and are thought to hold major implications for development, the environment, food safety and trade—both positive and negative. A close analysis of developments and implications of the high value beef chains in China, however reveals a less positive picture of high value agrifood chains than is usually reported in other industries and countries.

For instance, the attention given to the modernisation of retailing, especially supermarkets, is based on the assumption that they tightly control supply chains back to production level. There is little evidence of this control in China with the possible exceptions of pork and some fruit. For the enormous number of other agricultural products like beef that make up a small part of supermarket sales, it is simply beyond the capacity of supermarkets to successfully establish direct contracts with very large numbers of household producers and, in each case, to inspect and assess extension services, production inputs, production practices and outputs, to keep accounts and make payments, and to organise transport. Rather, these services are organised through and/or provided by processors or local government. Most often, however, supermarkets simply buy food on the open market and therefore do not directly manage these services, especially back to the production stage. The growth in the number of supermarkets in China should not, therefore, be assumed to hold significants implications for small-holders.

Compared to retailing, policy makers in China place far more attention on processing enterprises which, it is envisioned, will act as the "dragon heads" to modernise Chinese agriculture, help develop local economies and lead households to prosperity. The attention has resulted in the development of some impressive systems in the livestock sector, especially in the pig, poultry and dairy industries. However, a close analysis of the beef industry also reveals endemic over-capacity and unviability in the modern processing sector with corresponding detrimental effects.

Given the over-crowded nature of the modern processing sector relative to the small size of the premium market, processors are under enormous pressure to cut costs, which they do in various ways. First, capital costs and taxes are reduced by seeking preferential policies from government. Second, many abattoirs attempt to enter high value markets at low cost by marketing product that is not objectively graded, and can be subject to practices like false advertising, fake marbling and water injection. Third, while wages for slaughter workers in modern abattoirs are notionally on par with semi-skilled labour in other industries, the severe financial pressures faced by most large abattoirs means that they are under pressure to reduce labour costs. Abattoir employees are therefore employed on a temporary basis, only for the months of the year or the days of the week in which slaughter takes place. While temporary employment is common in the slaughter sectors of other countries, it is more pronounced in China because of endemic under utilisation in modern enterprises and the high level of seasonality in cattle, sheep and goat slaughter sector in China (due to seasonal consumption patterns and turnoff in pastoral systems). The fourth and most important means for abattoirs to reduce costs is in raw material procurement, especially through the downgrading of product (*yaji, yajia*). This factor helps explain why households are reluctant to enter into contracts for the supply of cattle to abattoirs and that abattoirs regularly complain about shortages of supply. Similar issues are reported for other agricultural industries in China.

These cost-cutting measures entail a range of negative impacts for government (in the form of wasted subsidies), consumers (fake product), workers (insecure employment) and producers (downgraded product and prices). Furthermore policy support to modern abattoirs in the conditions of an undeveloped premium beef market means that that the large modern abattoirs will sell a significant proportion of their product on to mid and even low value markets (at a loss) which effectively crowds out smaller slaughter operations. That is, while measures to build modern agricultural structures have been undertaken to benefit various sectors of society, these measures have been largely counter-productive from a public good perspective. The conclusion that follows from a circumspect analysis of modern agribusiness structures is that support can not be justified and should be reduced or eliminated completely. Some areas where these subsidies could be diverted and more effectively used—in training, standards and market development—have been mentioned above.

While analysis in this study identifies several areas of concern in pursuing agricultural modernisation through large, high cost and high value structures, it also finds that some of the concerns expressed in the literature about these structures are not applicable, at least in the case of beef. For example, it is sometimes argued that the potential contribution of modern supply chains to rural development will be limited or uneven because they exclude smaller scale producers. This is not the case in the beef industry for several reasons. First, the reason why contracts are not taken up on a widespread scale by small, poor households in the beef industry is not because they are too technically demanding or because they are not offered, but rather because of misconduct and a lack of trust in the contractual arrangements. Thus, the major question is not so much one of the inclusiveness of the contract systems, but how viable and well governed those systems are. Second, while most contracted households are specialised fattening units, these specialised households buy feeders from uncontracted unspecialised cow–calf households. As is the case with other industries with multiple stages of primary production (such as pigs), these uncontracted unspecialised households can be seen as being involved in, and potentially benefiting from, contract systems, albeit indirectly through the derived demand for their feeder cattle. However, this will occur only if the fattening operations that buy calves from the unspecialised households benefit from their contracts with the abattoirs.

With improved governance and conduct, contract agriculture and modern supply chains may well expand to gain a significant place in some agricultural industries, especially in higher value agrifood chains. In Chinese terms, households will become the "first workshop" in the "production line" of "industrialised agriculture". In doing so, households will increasingly forgo decision making and management functions. There are many cases where households simply contract out their land or where ownership of the livestock remains with enterprises while the household contribution is confined to the provision of labour. The mass uptake of such systems may be suited to many households. However, given China's turbulent modern history and the re-emergence of large inequalities in rural areas, the viability and governance of the contract systems will be critical.

Production

Another concern raised in the literature is that large scale modern production structures will squeeze out smaller producer households, regardless of whether they are contracted or not. This concern is also unfounded in the case of beef for a range of reasons, each of which also act as an argument against the support and encouragement of large scale modern production structures.

First, large production structures such as feedlots have higher (infrastructure, capital and labour) cost structures than specialised fattening households, which are not offset by scale or technical efficiencies. Despite the higher transaction costs of contracting large numbers of households, enterprises seek inputs at a lower price from labour intensive households than from larger capital intensive feedlots. This is especially the case if the households undervalue inputs (like labour) and are organised and supported by local government. Second, any growth that does occur in the feedlot sector would occur from a very low base and, given the underlying economics, would be slow. The growth would occur only in niche markets, namely in the production of very heavy cattle for premium beef markets. Third, even if feedlots were to increase their relative importance over the longer term, they usually finish cattle for only 90 to 120 days in order to increase control over quality and continuity of supply. Thus, feedlots still source most of their cattle (intermediate feeders) from household fattening operations. The problems of large scale agriculture crowding out household producers also appear remote in products such as dairy and (to a lesser extent) pigs, which are still mainly sourced from specialised households (although the "melamine incident" in the dairy industry has meant that dairy processors prefer to buy from larger dairies). Some differences, however, may apply in the broiler and layer sectors where large producers with scale efficiencies are vertically linked to hatcheries, feed and input suppliers, and where disease control is a major issue.

10.4 Concluding comments on policy recommendations

China faces major challenges in modernising its agricultural sector, especially in an environment dominated by small household-based industry participants, and especially given pressures to increase their incomes. However, accelerated modernisation through the development of high value, large scale modern structures and interventionist policies is not the way to meet this challenge. This study advocates a modernisation strategy based around developing mid value markets and agrifood chains. In this sense, Chinese agriculture needs to undergo an "evolution" rather than a "revolution". This general finding appears to be less relevant to industries such as poultry where technical and scale efficiencies are major factors in the viability and food safety of the systems. However, the principles that apply to poultry should not be applied to other livestock industries like beef and sheep meat that are more suited to small and mid scale structures in China into the medium future at least. Other industries such as pork and dairy will lie on other parts of the continuum. As an indication of the generalisability of this study, many of the findings from this study of the beef industry are closely aligned to the findings of Ke (2008, p.55) on the broader Chinese livestock sector, especially the pork industry.

The pursuit of a "middle line" agricultural modernisation strategy for the beef industry is not without problems. However, these problems can be addressed through carefully considered and refined policy and institutional settings. There are differences in this regard to the seemingly intractable problems involved in a focus on other value chains. Low value agrifood chains provide limited opportunity to increase incomes for producers and to provide safe meat to consumers, and China can not afford to take a "hands off", unregulated approach toward the low value agribusiness sector. At the same time, modernisation through the development of high value agrifood chains is limited by the small market and limited growth prospects, and measures to force the pace of industry growth and modernisation through "interventionist" policies have been largely counter-productive. Rather, attention now needs to turn toward the development of mid value markets and agrifood chains through the adoption of standards, measures to increase resource mobilisation, industry coordination and market development. These types of "market facilitating" policies are the key tools in China's pursuit of sustainable agricultural and rural development objectives into the future.

10.5 Further research

The preceding discussion highlights issues that will impact heavily on the future development of the beef industry and its participants. These issues have been examined in considerable detail in this study and in related publications by the author and his colleagues. However the dynamic and multi-dimensional nature of the issues provides scope for further analysis. Four issues in particular are explicitly identified for further research.

The first issue relates to change in the spatial development of the beef industry in the context of broader economic development in China. Increasing opportunity costs of labour in more developed agricultural areas of eastern China mean that small-scale cattle production—cow–calf production in particular—is becoming an increasingly unattractive economic activity for households in these areas. The subsequent decline in cattle supply, together with increasing wage rates and better labour conditions, also increases costs for abattoirs located in more developed parts of China. Higher cost structures in eastern China raise the prospect of multiple sectors of the beef industry moving westward, potentially introducing much needed development opportunities for the western China region. However, lower labour costs in western China must be weighed up against higher costs associated with greater distances to larger and higher value beef markets, especially cold storage and transport costs. Budgeting processes used in this study enable the estimation of the relative costs, revenues and returns for different categories of industry actors, in different geographical areas of China. However, insufficient data from producers and slaughter units in western China were collected and calibrated to be confidently incorporated into the models. While collecting and calibrating such data would be a resource-intensive process, it would provide important insights into how industry and rural development might and should develop into the future.

The discussion above also highlighted the implications of changing regulations toward the processing sector, slaughter units in particular. In many rural industries in China, the government is regulating and supporting the development of larger scale processing structures as a means of addressing increasingly urgent food safety and environmental problems. While larger, more centralised and more "standardised" processing structures can reduce inspection and monitoring costs for government, it can also preclude smaller rural actors from industry participation. Analysis in this study showed that a continuum of small to medium to large scale structures exists in the cattle slaughter sector each of which provide a series of policy options for policy makers. That is, each of these structures involve different trade offs between lower inspection costs and "safer" production structures on the one hand and participation from small scale operators and employment impacts on the other. The analysis in this study stopped short of quantifying the costs to government vis-à-vis the costs for small holder participation for each type of slaughter structure and regulatory regime. Such an analysis would help policy-makers assess policy options for the cattle slaughter sector and, given parallels with other rural industries, might prove instructive for other sectors of the rural economy.

A third and related issue is the assessment of agro-industrial systems and dragon head enterprises—including modern abattoirs—that are designed to "lead' households into higher value markets. As a critical juncture in the supply chain, it is imperative that dragon head enterprises are both commercially viable and also pass price-grade differentials back to suppliers. This study examined the systematic reasons why few modern abattoirs in China are commercially viable or benefit household suppliers. As dragon head abattoirs are established under agro-industrial programs using preferential policies and subsidies, the programs entail high opportunity costs and perverse impacts, especially on households integrated with the systems and enterprises. Thus an assessment of the cost-effectiveness of programs that support agro-industrial systems and dragon head enterprises is likely to have a strong positive social impact. Gathering commercially and politically sensitive data on the systems and enterprises is challenging, but as they are usually established under government or international programs, such assessments should be rigorously incorporated into the design, establishment and operational stages of the programs.

A fourth public policy issue that arose repeatedly throughout the study relates to standards applied to the marketing of both beef and cattle. The development—but more importantly the enforcement—of quality and food safety standards can be costly and, depending on the political economy of how these costs are distributed, must be borne by government, industry and/or consumers. However, these programs also deliver benefits to industry, industry participants including households, and consumers. A detailed cost-benefit analysis of the development and enforcement of different types of marketing standards was beyond the scope of this study. Such an analysis would provide policy makers with important information on which to develop marketing systems for the cattle and beef industry as well as other food and fibre industries in China.

Bibliography

Adelman (2001) Fallacies in Development Theory and Their Implications for Policy. In: Meier, G.M. and Stiglitz, J.E. (eds) *Frontiers of Development Economics: The Future in Perspective*. World Bank, New York.

Adelman, I. (1984) Beyond Export-Led Growth. *World Development* 12, 9, 937–949.

All China Data Centre, China Data Online, All China Marketing Research. Michigan State University.

Anonymous (2002) Muslims in Ningxia Benefit from Boom Trade, in *People's Daily*, September 25.

Anonymous (2005) Gansusheng Jiaxu Tuzai Guanli Banfa Jiang Shishe (Livestock Slaughter Management Regulations Will Be Implemented in Gansu Province), in *Gansu Nongmin Ribao (Gansu Peasant's Daily)*, 18 November 2005.

Ashley, C. and Maxwell, S. (2001) Rethinking Rural Development. *Development Policy Review* 19, 4, 395–425.

Barnard, N.D. (1999) *World Bank Versus World Health*, Physicians Committee for Responsible Medicine. Available at: www.pcrm.org/health/Commentary/commentary0001.html.

Barrett, C.B., Reardon, T. and Webb, P. (2001) Nonfarm Income Diversification and Livelihoods in Rural Africa: Concepts, Dynamics, and Policy Implications. *Food Policy* 29 4, 315–331.

Bertalanffy, L. (1976) *General System Theory: Foundations, Development, Applications*. George Brazillcr, New York.

Bi, X. (2007) *Who Is Benefiting from China's Livestock Boom?*, Western Educational Research Activity 101 Annual Meeting, Shanghai, 12–13 July.

Birthal, P.S., Joshi, P.K. and Gulati, A. (2005) *Vertical Coordination in High-Value Commodities*. MTID Discussion Papers, No. 85, International Food Policy Research Institute.

Blecher, M. and Shue, V. (2001) Into Leather: State-Led Development and the Private Sector in Xinji. *The China Quarterly* 166, 368–393.

Brown, C.G., Longworth, J.W. and Waldron, S.A. (2002a) Food Safety and Development of the Beef Industry in China. *Food Policy* 27, 3, 269–284.

Brown, C.G., Longworth, J.W. and Waldron, S.A. (2002b) *Regionalisation and Integration in China: Lessons from the Transformation of the Beef Industry*. Ashgate, Aldershot.

Brown, C.G., Waldron, S.A. and Longworth, J.W. (2005) *Modernizing China's Industries: Lessons from Wool and Wool Textiles*. Edward Elgar, Cheltenham.

Brown, C.G., Waldron, S.A. and Longworth, J.W. (2008) *Sustainable Development in Western China: Managing People, Livestock and Grasslands in Pastoral Areas*. Edward Elgar, Cheltenham.

Brown, C.G., Waldron, S.A., Longworth, J.W. and Zhao, Y.T. (2005) *Building Economic Decision-Making Capabilities of Chinese Wool Textile Mills*. ACIAR Technical Reports No. 60, Australian Centre for International Agricultural Research, Canberra.

Byrd, W.A. and Lin, Q. (1990) *China's Rural Industry: Structure, Development, and Reform*. Oxford University Press, Oxford.

Central Committee of Chinese Communist Party (1998) Decisions of Key Issues on Agriculture and Rural Work.

Chambers, R. and Conway, G. (1992) Sustainable Rural Livelihoods: Practical Concepts for Thetwenty-First Century. Vol. IDS Discussion paper 296, Institute of Development Studies, University of Sussex, Brighton.

Chang, G.H. (2002) The Cause and Cure of China's Widening Income Disparity. *China Economic Review* 13, 4, 335–340.

Chen, J., Carter, C., Rozelle, S. and Huang, J. (2005) *Market Emergence and the Rise and Fall of Backyard Hog Production in China*. CESP Working Paper, Centre for Environmental Sciences and Policy, Stanford University.

China News Service (2006) Ministry of Agriculture Launches Plan for Rural Market Info Network. September 30.

China.org.cn(2006)AnimalHusbandryLawEffectiveJuly1,June27,2006.Availableat: http://www.china.org.cn/government/central_government/2006–06/27/content_1172817.htm.

Coase, R. (1937) The Nature of the Firm. *Economica* 4, 386–405.

Cook, M.L. and Chaddad, F.R. (2000) Agroindustrialization of the Global Agrifood Economy: Bridging Development Economics and Agribusiness Research. *Agricultural Economics* 23, 3, 207–218.

Creswell, J.W. and Clark, V.L.P. (2003) *Designing and Conducting Mixed Methods Research*. Sage Publications, Thousand Oaks.

De Brauw, A., Huang, J., Rozelle, S., Zhang, L. and Zhang, Y. (2002) The Evolution of China's Rural Labor Markets During the Reforms. *Journal of Comparative Economics* 30, 2, 329–353.

De Haen, H., Stamoulis, S., Shetty, P. and Pingali, P. (2003) The World Food Economy in the Twenty-First Century: Challenges for International Co-operation. *Development Policy Review* 21, 5/6, 683–696.

Deininger, K. and Jin, S. (2005) The Potential of Land Markets in the Process of Economic Development: Evidence from China. *Journal of Development Economics* 78, 1, 241–270.

Delgado, C., Rosegrant, M., Steinfeld, H., Ehui, S. and Courbois, C. (1999a) The Coming Livestock Revolution. *Choices*, (Fourth Quarter, Choices at the Millennium. A Special Issue), 40–44.

Delgado, C., Rosegrant, M., Steinfeld, H., Ehui, S. and Courbois, C. (1999b) *Livestock to 2020: The Next Food Revolution*. International Food Policy Research Institute, FAE Discussion Paper 28, International Food Policy Research Institute, Food and Agriculture Organization of the United Nations, and International Livestock Research Institute, Washington DC.

Delman, J. (2003) Cool Thinking? The Role of the State in Shaping China's Dairy Sector and Its Knowledge System. *China Information* 17, 2, 1–35.

Diao, X.S and Hazell, P. (2004) Exploring Market Opportunities for African Smallholders. 2020 Africa Conference Brief 6, International Food Policy Research Institute, Washington DC.

Diao, X.S., Hazell, P., Resnick, D. and Thurlow, J. (2006) The Role of Agriculture in Development: Implications for Sub-Saharan Africa. DSGD Discussion Paper No. 29, International Food Policy Research Institute.

Dick, B. and Swepson, P. (1997) *Action Research Frequently Asked Questions.* Available at: Available at http://www.scu.edu.au/schools/gcm/ar/arp/arfaq.

Dopfer, K. (ed) (2001) *Evolutionary Economics: Program and Scope.* Kluwer Academic Publishers, Dordrecht.

Dorward, A. and Kydd, J. (2005) Making Agricultural Market Systems Work for the Poor: Promoting Effective, Efficient and Accessible Coordination and Exchange. Proceedings of Asian Development Bank Joint workshop on Making Markets Work for the Poor, Manila, 15–16 February 2005.

Du, R.S. (2006) The Course of China's Rural Reform. International Food Policy Research Institute occasional paper.

Editorial Board of China Agriculture Yearbook (ed) (2006) *China Agriculture Yearbook.* China Agriculture Publishing House, Beijing.

Editorial Board of the China Animal Husbandry Yearbook (2002) *China Animal Husbandry Yearbook.* China Agriculture Publishing House, Beijing.

Editorial Board of the China Township and Village Enterprise Yearbook (2006) *China Township and Village Enterprise Yearbook.* China Agriculture Publishing House, Beijing.

Ellis, F. (2000a) The Determinants of Rural Livelihood Diversification in Developing Countries. *Journal of Agricultural Economics* 51, 2, 289–302.

Ellis, F. (2000b) *Rural Livelihoods and Diversity in Developing Countries.* Oxford University Press, Oxford.

Ellis, F. and Biggs, S. (2001) Evolving Themes in Rural Development 1950s-2000s. *Development Policy Review* 19, 4, 437–448.

Ellis, F. and Harris, N. (2004) New Thinking About Urban and Rural Development. Proceedings of Keynote Paper for Department ofr International Development Sustainable Development Retreat.

Fabiosa, J.F., Hu, D.H. and Fang, C. (2005) *A Case Study of China's Commercial Pork Value Chain.* MATRIC Research Paper 05-MRP 11, Midwest Agribusiness Trade Research and Information Center, Iowa State University.

FAO (2001) Weaknesses in the Livestock Industries of Asia: Results from a Survey of Livestock Policy Makers and Researchers. Food and Agriculture Organisation of the United Nations, Bangkok.

FAO (2007a) Challenges of Agribusiness and Agro-Industries Development. In *Twentieth Session Commitee on Agriculture*, Food and Agriculture Organization of the United Nations.

FAO (2007b) *Pro-Poor Review and Analysis of Livestock Related Policies* The Pro-Poor Livestock Policy Initiative of the Food and Agriculture Organisation of the United Nations.

Fei, J.C. and Ranis, G. (1961) A Theory of Economic Development. *American Economic Review* 51, 4, 533–565.

Finch, B. and Longworth, J.W. (2000) *The Hong Kong Beef Market: Opportunities for Australia*, Information Series Report No. QI00011, The State of Queensland Department of Primary Industries, Brisbane.

Friedman, M. (1953) *Essays in Positive Economics.* University of Chicago Press, Chicago.

Fuller, F., Hayes, D. and Smith, D. (2000) Reconciling Chinese Meat Production and Consumption Data. *Economic Development and Cultural Change* 49, 1, 23–43.

Fuller, F., Hu, D.H., Huang, J.K. and Hayes, D.J. (2001) Livestock Production and Feed Use by Rural Households in China: A Survey Report. Staff Report 01 SR-96, Food and Agricultural Policy Research Institute, Iowa State University.

Gale, F. and Huang, K. (2007) *Demand for Food Quantity and Quality in China.* Economic Research Report Number 32, Economic Research Service, United States Department of Agriculture, January 2007.

Gale, F. (2005) *Commercialization of Food Consumption in Rural China.* Economic Research Report Number 8, Economic Research Service, United States Department of Agriculture, July 2005.

Glaser, B.G. (1994) *Basics of Grounded Theory Analysis: Emergence Versus Forcing*. Sociology Press, Mill Valley.

Glover, D. and Kusterer, K. (1990) *Small Smallholders, Big Business: Contract Farming and Rural Development*. Macmillan, London.

Goldman, A. (2000) Supermarkets in China: The Case of Shanghai. *The International Review of Retail Distribution and Consumer Research*, 1, 23, (Reprinted in L. Sparks and A. Findlay eds. (2001) *Retailing: Critical Concepts*, Routledge).

Goldman, A. (2003) Evaluating the Modernization of the Chinese Retail System. Working Paper, School of Business Administration, The Hebrew University of Jerusalem.

Goldman, A., Krider, R. and Ramaswami, S. (1996) *Factors Impeding Market Share Growth of Supermarkets: Food Retail Modernization in Hong Kong*. Working Paper Series No. MKTG 96.082, Business School, Hong Kong University.

Goldman, A. and Vanhonacker, W. (2006) The Food Retail System in China: Strategic Dilemmas and Lessons for Retail Internationalization/Modernization In *Globalizing Retail Conference*, University of Surrey.

Gong, W. (2007) China's Developing Beef Industry: A Supply Chain Management Analysis. Dissertation of the Faculty of Rural Management, University of Sydney.

Guo, H.D., Jolly, R.W. and Zhu, J.H. (2007) Contract Farming in China: Perspectives of Farm Households and Agribusiness Firms. *Comparative Economic Studies* 49, 2, 285–312.

Haggblade, S., Hazell, P. and Brown, J. (1989) Farm-Nonfarm Linkages in Rural Sub-Saharan Africa. *World Development* 17, 8, 1173–1202.

Hayami, Y. and Otsuka, K. (1993) *The Economics of Contract Choice*. Oxford University Press, Oxford.

Hazell, P. and Haggblade, S. (1991) Rural-Urban Growth Linkages in India. *Indian Journal of Agricultural Economics* 46, 4, 512–529.

Hazell, P., Poulton, C. and Wiggins, S. (2006) The Future of Small Farms. In *Paper presented at the annual conference of the International Association of Agricultural Economists*, Gold Coast, August 12–18, 2006.

Hazell, P. and Roell, A. (1983) *Rural Growth Linkages: Household Expenditure Patterns in Malaysia and Nigeria*. Research Reports 41, International Food Policy Research Institute, Washington DC.

Heron, J. (1971) Experience and Method: An Inquiry into the Concept of Experiential Research. In *Human Potential Research Project*, University of Surrey.

Herriot, R.E. and Firestone, W.A. (1983) Multisite Qualitative Policy Research: Optimizing Description and Generalizability. *Educational Researcher* 12, 3, 14–19.

Hoffmann, D. (1999) *Asian Livestock to the Year 2000 and Beyond*. Food and Agriculture Organization of the United Nations, Working Paper Series 1/2.

Hu, D.H, Cai, H. and van Gelder, R.A. (1998) *Beef Consumption and the Beef Marketing Chains in China*. Discussion Paper 04/98, School of Natural and Rural Systems Management, The University of Queensland.

Hu, D.H., Reardon, T., Rozelle, S., Timmer, P. and Wang, H. (2004) The Emergence of Supermarkets with Chinese Characteristics: Challenges and Opportunities for China's Agricultural Development. *Development Policy Review* 22, 5, 557–586.

Huang, J.K. and Rozelle, S. (2006) The Emergence of Agricultural Commodity Markets in China. *China Economic Review* 17, 3, 266–280.

Huang, J.K., Dong, X.X., Wu, Y.H., Zhi, H.Y., Nui, X.F., Huang, Z.R. and Rozelle, S. (2007) *Restructuring Agrifood Markets in China: The Horticulture Sector (a)*. Regoverning Markets Agrifood Sector Studies, International Institute for Environment and Development, London.

Huang, J.K., Wu, Y.H., Zhi, H.Y. and Rozelle, S. (2008) Small holder incomes, food safety and producing, and marketing China's fruit. *Review of Agricultural Economics*, 30, 469–479.

IFAD (2001) *Rural Poverty Report 2001: The Challenge of Ending Rural Poverty*. Oxford University Press, Oxford.

ILRI (2007) *Livestock Asia: A Strategy and Action Plan for Research for Poverty Reduction*. International Livestock Research Institute. Available at www.ilri.org/ILRIPubAware/Uploaded%20Files/Action%20Plan%2019%20Nov%20.pdf.

Irz, X., Lin, L., Thirtle, C. and Wiggins, S. (2001) Agricultural Growth and Poverty Alleviation. *Development Policy Review* 19, 4, 445–467.

Jiang, H.M. (1995) *Youzhi Niurou Shengchan Jishu (Quality Beef Production Techniques)*. China Agriculture Publishing House, Beijing.

Johnston, B.F. and Kilby, P. (1975) *Agriculture and Structural Transformation*. Oxford University Press, New York.

Johnston, D.G. and Mellor, J.W. (1961) The Role of Agriculture in Economic Development. *American Economic Review* 51, 4, 566–593.

Joskow, P.L. (2005) Vertical Integration. In: Menard, C. and Shirley, M.M. (eds) *Handbook of New Institutional Economics*. Dordrecht, Springer.

Ke, B.S. (2001) Recent Development in the Livestock Sector in China and Changes in Livestock/Feed Relationship. In *Agricultural and Natural Resource Economics Discussion Paper*, No. 01/2001, School of Natural and Rural Systems Management, The University of Queensland.

Ke, B.S. (2002) *Perspectives and Strategies for the Livestock Sector in China in the Next Three Decades*. Livestock Policy Discussion Paper No.27, in "Perspectives and Strategies for the Asian Livestock Sector in the next three decades". Food and Agriculture Organization of the United Nations.

Ke, B.S. (2008) *Gongye Fanbu Nongye De Lilun Yu Shixian Yanjiu (Research on the Theory and Practice of Reverse Transfers from Industry to Agriculture)*. People's Publishing House, Beijing.

Key, N. and Runsten, D. (1999) Contract Farming, Smallholders, and Rural Development in Latin America: The Organization of Agroprocessing Firms and the Scale of Outgrower Production. *World Development* 27 2, 381–401.

Knight, J. and Song, L.N. (2003) Chinese Peasant Choices: Migration, Rural Industry, or Farming. *Oxford Development Studies* 31, 2, 123–148.

Kolb, D.A. (1984) *Experiential Learning: Experience as the Source of Learning and Development*. Prentice-Hall, Englewood Cliffs.

Kumar, A. (1994) *China: Internal Market Development and Regulation*. The World Bank, Washington DC, 248 pp.

Lewin, K. (1946) Action Research and Minority Problems. *Journal of Social Issues* 2, 4, 34–46.

Li, S. and Li, S. (2000) Guanxi: Just Like Traffic Cops. *China Economic Quarterly* 4, 1.

Li, Z.Q., Feng, Y.Q., Dong, X.X. and Han, S.W. (2008) Rupin Jiage Xingcheng Ji Lirun Fenpei De Diaocha Baogao (Investigation of Dairy Price Transmission and Profit Distribution of Dairy Farms). *Nongye Zhanwang (Agricultural Outlook)* 4, 5, 20–22.

Lieberthal, K. and Oksenberg, M. (1988) *Policy Making in China: Leaders, Structures, and Processes*. Princeton University Press, Princeton.

Lin, J.Y. (1992) Rural Reforms and Agricultural Growth in China. *The American Economic Review* 82, 1, 34–51.

Lin, X.J. (1997) Zhongguo Xumuye De Tezheng Muqian Fazhan Jieduan Ji Zhengfu De Zuoyong (on Characteristics of China's Animal Husbandry in Terms of Its Current Development Phase and Roles of Government). *Zhongguo Nongcun Jingji (Chinese Rural Economy)*, 9, 66–69.

Lin, X.J. (1998) *Xumuye Chanyehua Wenti Yanjiu (Research on Issues of Integration of the Livestock Sector)*. China Agricultural Scientech Publishing House, Beijing.

Lindblom, C.E. (1977) *Politics and Markets: The World Economic System*. Basic Books, New York.

Lingohr, S. (2007) Rural Households, Dragon Heads and Associations: A Case Study of Sweet Potato Processing in Sichuan Province. *The China Quarterly* 192, 898–914.

Liu, H.B., Parton, K., Zhou, Z.Y. and Cox, R. (2007) Meat Consumption in the Home in China: An Empirical Study. Paper Presented at the International Agricultural Trade Research Consortium, China's Agriculture Trade: Issues and Prospects, Beijing. July 8–9, 2007.

Liu, Y.M. (1998) The Impacts of Developing Beef and Cattle Industry on Household Incomes and Labour Utilisation. In: Liu, Y. (ed) *Zhongguo Rouniuye Jingji Wenti Yanjiu (Economic Aspects of the Beef and Cattle Industry in China)*. Contemporary China Publishing House, Beijing, pp. 260–269.

Liu, Y.M. (2008) Aozhou Xuezhe Dui Woguo Niurou Yangzhiye De Jingji Xiaoyi Fenxi Ji Liugei Women De Qushi (Comments from Australian Scholars on the Economic Efficiency of China's Beef Cattle Industry). In *Proceedings of the Third China National Cattle Industry Development Conference*, China Publishing House, Muling, Heilongjiang.

Liu, Y.M. (ed) (2000) *Zhongguo Rouniuye Jingji Wenti Yanjiu (Economic Aspects of the Beef and Cattle Industry in China)*. Contemporary China Publishing House, Beijing.

Longworth, J.W. and Brown, C.G. (1995) *Agribusiness Reforms in China: The Case of Wool*. CAB International, Wallingford.

Longworth, J.W., Brown, C.G. and Waldron, S.A. (2001) *Beef in China: Agribusiness Opportunities and Challenges*. University of Queensland Press, St Lucia.

Longworth, J.W., Brown, C.G. and Waldron, S.A. (2003) *Zhongguo Rouniuye De Shangun Jiyu Yu Tiaozhan (Beef in China: Agribusiness Opportunities and Challenges)*. China Agriculture Publishing House, Beijing.

Longworth, J.W. and Williamson, G.J. (1993) *China's Pastoral Region: Sheep and Wool, Minority Nationalities, Rangeland Degradation and Sustainable Development*. CAB International, Wallingford.

Lu, H.L. (2007) The Role of Guanxi in Buyer-Seller Relationships in China: A Survey of Vegetable Supply Chains in Jiangsu Province. PhD Dissertation from Wageningen University.

Lu, X.P. (1994) Development of China's Beef Industry and Market Opportunities for the Australian Industry. Masters Dissertation Thesis, Department of Agriculture, The University of Queensland, St Lucia.

Ma, H., Rae, A., Huang, J.K. and Rozelle, S. (2004) Chinese Animal Product Consumption in the 1990s. *The Australian Journal of Agricultural and Resource Economics* 48, 8, 569–590.

Ma, H.Y., Huang, J.K., Fuller, F. and Rozelle, S. (2006) Getting Rich and Eating Out: Consumption of Food Away from Home in Urban China. *Canadian Journal of Agricultural Economics* 54, 1, 101–119.

Ma, H.Y., Huang, J.K. and Rozelle, S. (2005) Reassessing China's Livestock Statistics: Analyzing the Discrepancies and Creating New Data Series. *Economic Development and Cultural Change* 52, 117–131.

MacIntyre, A.J. (1989) Corporatism, Control and Political Change in 'New Order' Indonesia. In: May and O'Malley (eds) *Observing Change in Asia*. Crawford House Press, Bathurst.

McMichael, P. (2001) The Impact of Globalisation, Free Trade and Technology on Food and Nutrition in the New Millennium. *Proceedings of the Nutrition Society* 60, 2, 215–220.

Mellor, J.W. (1976) *The New Economics of Growth*. Cornell University Press, Ithaca.

Meng, Z. (2007) Are Chinese Health Concerned Food Consumers? A Panel Study of Household Food Demand in China from 1989 to 2004. Paper Presented at International Agricultural Trade Research Consortium, China's Agriculture Trade: Issues and Prospects, Beijing. July 8–9, 2007.

Ministry of Agriculture (2001) *Opinion of Accelerating Livestock Development*. Joint Release from the Ministry of Agriculture and the State Council, 20 October.

Ministry of Agriculture (2004) *China Agriculture Yearbook*. China Agriculture Publishing House, Beijing.

Ministry of Agriculture (2005) *China Agricultural Development Report*. China Agriculture Publishing House, Beijing.

Ministry of Agriculture (2006) *China Agriculture Yearbook*. China Agriculture Publishing House, Beijing.

Ministry of Commerce (2006) Guonei Maoyi "Shiyiwu" Fazhan Mubiao Yijing Queding (the Development Goals for the 11th Five Year Plan for Internal Trade Has Been Established).

Minot, N. (2005) Do Small-Scale Producers Gain from Supply Chain Coordination? The Case of High-Value Agriculture in Asia. In *Workshop on "Linking Small-Scale Producers to Markets: Old and New Challenges"*, The World Bank, 15 December 2005.

Nan, J. and Han, Y.J. (2008) Zhurou Jiage Xingcheng Gucheng Ji Lirun Fenpei Qingkuang Diaoyan Baogao (Invgestigation of Pork Price Transmission and and Benefit Distribution of Pork Chain). *Nongye Zhanwang (Agricultural Outlook)* 4, 5, 13–16.

National Bureau of Statistics (Various years) *China Statistical Yearbook*. China Statistics Publishing House, Beijing.

National Office for Rural Household Surveys (1998) A Comparative Study of Production Efficiency and Profitability of Various Animal Raising Practices. *Reference Materials for Rural Economic Research* 9, 20–25.

Nie, P. (2006) Ministry Releases 5-Year Trade Plan, in *China Daily/Xinhua, reported on gov.cn*, Friday, August 25, 2006 Available at: http://english.gov.cn/2006–08/25/content_369761.htm.

Niu, R.F. (2002) Zhongguo Nongye Chanyehua Jingying De Fazhan Tedian Yu Fangxiang (Development Characteristics and Directions in Agricultural Vertical Integration Management in China). *Zhongguo Nongye Jingji (China Agricultural Economy)* 5, 4–8.

Niu, R.F. and Xia, Y. (2000) *Nongye Chanyehua Jingying De Zuzhi Fangshi He Yunxing Jizhi (Organisational Patterns and Operation Mechanism of Agricultural Vertical Integration Management)*. Beijing University Press, Beijing.

OECD (2000) *The Agro-Food Processing Sector in China: Developments and Challenges*. Organisation for Economic Co-operation and Development, Paris.

Oi, J.C. (1999) *Rural China Takes Off: Institutional Foundations of Economic Reform*. University of California Press, Berkeley.

Organisation for Economic Co-operation and Development (2000) *The Agro-Food Processing Sector in China: Developments and Challenges*. OECD, Paris.

Pan, Y.G. (1998) *New Pattern of the Circulation System for Livestock Products in China*. Ministry of Agriculture, Government Report.

Patrick, I. (2004) *Contract Farming in Indonesia: Smallholders and Agribusiness Working Together*. ACIAR Technical Reports No. 54, Australian Centre for International Agricultural Research, Canberra.

Pingali, P. (2007) Westernization of Asian Diets and the Transformation of Food Systems: Implications for Research and Policy. *Food Policy* 32, 3, 281–298.

Poulton, C. and Dorward, A. (2003) *The Role of Market Based Economic Development in Strengthening Food Security*. Forum for Food Security in Southern Africa.

Rae, A.N (1997) Changing Food Consumption Patterns in East Asia: Implications of the Trend Towards Livestock Products. *Agribusiness* 13, 1, 33–44.

Rae, A.N (1998) The Effects of Expenditure Growth and Urbanisation on Food Consumption in East Asia: A Note on Animal Products. *Agricultural Economics* 18, 3, 291–299.

Rae, A.N (2008) China's Agriculture, Smallholders and Trade: Driven by the Livestock Revolution? *Australian Journal of Agricultural and Resource Economics* 52, 3, 283–302.

Rae, A.N, Ma, H.Y, Huang, J.K and Rozelle, S. (2006) Livestock in China: Commodity-Specific Total Factor Productivity Decomposition Using New Panel Data. *American Journal of Agricultural Economics* 88, 3, 680–695.

Rae, A.N and Zhang, X.H. (2009) China's booming livestock industry: household income, specialization, and exit. *Agricultural Economics*, 40, pp. 603–616.

Ramaswami, B., Birthal, P.S. and Joshi, P.K. (2006) Efficiency and Distribution in Contract Farming: The Case of Indian Poultry Growers. In *MTID Discussion Papers*, International Food Policy Research Institute.

Rammohan, K.T. and Sundaresan, R. (2003) Socially Embedding the Commodity Chain: An Exercise in Relation to Coir Yarn Spinning in Southern India. *World Development* 31, 5, 903–923.

Ravallion, M. (2005) Externalities in Rural Development: Evidence for China. In: Ravi, K. and Venables, A.J. (eds) *Spatial Inequality and Development*. Oxford University Press, Oxford.

Reardon, T., Codron, J.M., Busch, L., Bingen, J. and Harris, C. (2001) Global Change in Agrifood Grades and Standards: Agribusiness Strategic Responses in Developing Countries. *International Food and Agribusiness Management Review* 2, 3/4, 421–435.

Reardon, T., Fall, A., Kelly, V., Delgado, C., Matlon, P., Hopkins, J. and Badiane, O. (1994) Is Income Diversification Agriculture-Led in the West African Semi-Arid Tropics? In: Atsain, A., Wangwe, S. and Drabek, A.G. (eds) *Economic Policy Experience in Africa*. African Economic Research Consortium, Nairobi.

Riethmuller, P.C. (2002) *Some Issues Associated with the Livestock Industries of the Asia-Pacific Region*. RAP Publication 2002/06, Food and Agriculture Organization of the United Nations.

Royal Tropical Institute (2006) *Chain Empowerment: Supporting African Farmers to Develop Markets*. Faida MaLi and International Institute of Rural Reconstruction.

Rozelle, S. (2008) China's Rural Economy and the Path Towards a Modern Industrial State. Paper presentated at the Consortium for Western China Development Studies, "Assessment of Western China Development and Future Prospects", Xi'an, July 22–24, 2008.

Rozelle, S., Huang, J. and Benziger, V. (2003) Continuity and Change in China's Rural Periodic Markets. *The China Journal* 49, 89–115.

Rozelle, S., Huang, J. and Zhang, L. (2002) Emerging Markets, Evolving Institutions, and the New Opportunities for Growth in China's Rural Economy. *China Economic Review* 13, 4, 345–353.

Schlosser, E. (2001) *Fast Food Nation: What the All-American Meal Is Doing to the World*. Allen Lane, London.

Schultz, T.W. (1964) *Transforming Traditional Agriculture*. Yale University Press, New Haven.

Senge, P.M., Roberts, C., Ross, R., Smith, B. and Kleiner, A. (1994) *The Fifth Discipline Fieldbook: Strategies and Tools for Building a Learning Organization*. Doubleday Currency, New York.

Shandong Animal Husbandry Bureau (2005) Shandong Xumuye "Shiyiwu" Fazhan Guihua (Development Plan for the Shandong Livestock Industry for the 11th Five Year Plan).

Shandong Industry and Commerce Bureau (2006) Shandong Industry and Commerce System. Government document provided to the authors by the Shandong Industry and Commerce Bureau.

Shandong Province Animal Husbandry Office (ed) (2004) *Quansheng Xumu Hezuoshe Jingji Zuzhi Jingyan Jiaoliuhui Dianxing Cailiao (Cooperative Economy Organisations in the Province: Compilation of Materials to Exchange Experiences).*

Shen, G., Chen, Y.C., Liu, Q.D. and Zhao, H. (2008) Zhongguo Niurou Chanye De Fazan Yu Tiaozheng (China's Beef Industry Development and Challenges). In: Zhongguo Xumuye Xiehui (China Animal Husbandry Associaition) (ed) *China's Cattle Industry Development*. China Agriculture Publishing House, Beijing.

Shen, M.G., Rozelle, S. and Zhang, L.Z. (2005) *Farmer's Professional Associations in Rural China: State Domination or New State-Society Partnerships?* FED Working Papers Series No.FE20050013.

Shi, X., Heerink, N. and Qu, F. (2007) Choices between Different Off-Farm Employment Sub-Categories: An Empirical Analysis for Jiangxi Province, China. *China Economic Review* 18, 4, 438–455.

Sicular, T. (1996) Redefining State, Plan and Market: China's Reforms in Agricultural Commerce. In: Walder, A.G. (ed) *China's Transitional Economy*. Oxford University Press, Oxford, pp. 58–84.

Simmons, P., Winters, P. and Patrick, I. (2005) An Analysis of Contract Farming in East Java, Bali and Lombok, Indonesia. *Agricultural Economics* 33, 513–525.

Simpson, J. (2003) Long-Term Projections of Livestock, Meat and Feedstuffs in China: Focus on Beef Production Potential. Proceedings of 2003 Annual Meeting of WCC-101: China as a Market and Competitor, Portland, Oregon, 17–18 April.

Simpson, J., Cheng, X. and Miyazaki, A. (1994) *China's Livestock and Related Agriculture: Projections to 2025*. CAB International, Wallingford.

Singer, H. (1979) Policy Implications of the Lima Target. *Industry and Development* 3, 17–23.

Singh, S. (2000) Theory and Practice of Contract Farming: A Review. *Journal of Social and Economic Development* 3, 2, 255–263.

Skinner, G.W. (1964) Marketing and Social Structure in Rural China: Part 1. *Journal of Asian Studies* 24, 1, 3–43.

Skinner, G.W. (1965a) Marketing and Social Structure in Rural China: Part 2. *Journal of Asian Studies* 24, 2, 195–228.

Skinner, G.W. (1965b) Marketing and Social Structure in Rural China: Part 3. *Journal of Asian Studies* 24, 3, 363–399.

Skinner, G.W. (1985) Rural Marketing in China: Repression and Revival. *China Quarterly* 103, 393–413.

Skyttner, L. (2006) *General Systems Theory: Problems, Perspective, Practice*. World Scientific Publishing Company, River Edge.

Smith, D.C. (2001) Chinese Live Cattle and Beef Marketing and Distribution Systems. PhD Dissertation Thesis, School of Natural and Rural Systems Management, The University of Queensland.

State Council (1999) Opinion of Accelerating Livestock Development.

State Council (2003) *Opinion of Accelerating Development Processing Industry of Agricultural Products*. State Council, 14 January.

Strauss, A.L. and Corbin, J. (1990) *Basics of Qualitative Research: Grounded Theory Procedures and Techniques*. Sage Publishing, Newbury Park.

The Study Group of the Chinese Association for the Promotion of International Agricultural Co-operation (1997) Wuguo Nongqu Yangniuye De Fazhan Ji Qishi (Development of Inspiration for the Cattle Industry in Agricultural Areas of China). *Agricultural Economics*, 2, 3–15.

Unger, J. and Chan, A. (1995) China, Corporatism, and the East Asian Model. *The Australian Journal of Chinese Affairs*, 33, 29–53.

USDA (1998) *Statistical Revisions Significantly Alter China's Livestock Production, Supply and Demand*, United States Department of Agriculture, Foreign Agricultural Service, October 23, 1998. Available at: www.fas.usda.gov/dlp2/circular/1998/98–10lp/chinarv4.htm.

USDA (2005) *China, Peoples Republic of, Agricultural Situation, Wholesale Markets and Price Reporting*. GAIN Report CH5009, United States Department of Agriculture, Foreign Agricultural Service.

USDA (2006) *Livestock and Poultry: World Market and Trade Circular Archives*. United States Department of Agriculture, Foreign Agricultural Service, March 2006.

USDA (2007) *Agricultural Projections to 2016*. United States Department of Agriculture, Foreign Agricultural Service.

Veeck, A. and Burns, A.C. (2005) Changing Tastes: The Adoption of New Food Choices in Post-Reform China. *Journal of Business Research* 58, 5, 644–652.

Waldron, S.A. (1999) *Models of Agro-Industrialisation in China: The Case of the Cattle and Beef Industry*. Working Paper 07/99, School of Natural and Rural Systems Management, University of Queensland. Available on www.nrsm.uq.edu.au/caeg.

Waldron, S.A. (2007) Transitions in Rural China from the Perspective of the Beef Industry and Its Participants. International Agricultural Trade Research Consortium, China's Agriculture Trade: Issues and Prospects, Beijing. July 8–9.

Waldron, S.A. (2008) Comments from Australian Scholars on the Economic Efficiency of China's Beef Cattle Industry. Invited Presentation at the Third China National Cattle Industry Development Conference. Muling, July 26–28.

Waldron, S.A. and Brown, C.G. (2006) State Sector Reform and Agriculture in China. *The China Quarterly* 186, 277–294.

Waldron, S.A., Brown, C.G. and Longworth, J.W. (2003a) *Rural Development in China: Insights from the Beef Industry*. Ashgate, Aldershot.

Waldron, S.A., Brown, C.G. and Longworth, J.W. (2003b) Transforming Lives with Livestock-Based Agriculture. In: Brown, A.G. (ed) *The Livestock Revolution: A Pathway from Poverty?* The ATSE Crawford Fund, Canberra, pp. 27–39.

Waldron, S.A., Brown, C.G., Longworth, J.W. and Zhang, C.G. (2007) *China's Livestock Revolution: Agribusiness and Policy Developments in the Sheepmeat Industry*. CAB International, Wallingford.

Waldron, S.A., Brown, C.G., Longworth, J.W., Zhang, C.G. and Liu, Y.M. (2004) Impact Assessment of ACIAR Project ANRE95/02 Analysis of Socio-Economic and Agribusiness Developments in the Chinese Cattle and Beef Industry for the Australian Centre for International Agricultural Research, Canberra.

Waldron, S.A., Brown, C.G. and Zhang, C.G. (2007) *Update on Developments in the Chinese Cattle and Beef Industry of Relevance to the Australian Industry*. Report prepared for Meat and Livestock Australia. Available at www.nrsm.uq.edu.au/caeg/, 260 pp.

Wall, W.D. and Williams, H.L. (1970) *Longitudinal Studies and the Social Sciences*. Heinemann Educational Books, London.

Wang, H.L., Dong, X.X., Rozelle, S., Huang, J.K. and Reardon, T. (2009) Producing and Procuring Horticultural Crops with Chinese Characteristics: The Case of Northern China. *World Development*, 37, 1791–1801.

Wang, J.M., Zhou, Z.Y. and Yang, J. (2004) How Much Animal Product Do the Chinese Consume: Empirical Evidence from Household Surveys. *Australasian Agribusiness Review* 12, Paper 4.

Wang, L. and Davis, J. (1992) The Development and Reform of the Rural Marketing System in China. In: Vermeer, E.B. (ed) *From Peasant to Entrepreneur: Growth and Change in Rural China*. Centre for Agricultural Publishing and Documentation, Wageningen, pp. 69–82.

Wang, X.J. and Zhang, W.X. (1994) An Economic Assessment on Extending Ammoniated Straw and Stalk. *China Feed*, 5.

Wang, Z., Mao, Y. and Gale, F. (2008) Chinese Consumer Demand for Food Safety Attributes in Milk Products. *Food Policy* 33, 1, 27–36.

Wang, Z.G. (2003) Consumers' Recognition of Food Safety and Their Decision Making for Consumption. *Chinese Rural Economy* 4, pp. 41–51.

Wang, Z.G. (2006) *Market, Food Safety and China Agricultural Development*. China Agricultural Science and Technology Press, Beijing.

Warning, M. and Key, N. (2002) The Social Performance and Distributional Consequences of Contract Farming: An Equilibrium Analysis of the Arachide De Bouche Program in Senegal. *World Development* 30, 2, 255–263.

Watson, A. (1988) The Reform of Agricultural Marketing in China since 1978. *The China Quarterly* 113, 1–28.

Webb, E.J., Campbell, D.T., Sechrest, L. and Schwartz, R. (2000) *Unobtrusive Measures*. Sage Publishing, Thousand Oaks.

Weinberger, K. and Lumpkin, T.A. (2005) *High Value Agricultural Products in Asia and the Pacific for Small-Holder Farmers: Trends, Opportunities and Research Priorities*, AVRDC working paper prepared for a conference titled "How Can the Poor Benefit from the Growing Markets for High Value Agricultural Products", Cali, Colombia, October 3–5.

Weiss, R.S. and Rein, M. (1970) The Evaluation of Broad-Aim Programs, Experimental Design, Its Difficulties and an Alternative. *Administrative Science Quarterly* 15, 97–109.

Wilkes, A. (2003) Using the Livelihoods Framework to Understand Agro-Pastoralist Livelihoods in Northwest Yunnan. Working Paper 2, Center for Biodiversity and Indigenous Knowledge Community Livelihoods Program

Williams, M. and May, T. (1996) *Introduction to the Philosophy of Social Research*. UCL Press, London.

World Bank (1999) *China—Smallholder Cattle Development Project: Project Appraisal Document*. Rural Development and Natural Resources Sector Unit, East Asia and Pacific Region.

World Bank (2001) *Livestock Development, the Environment, Poverty and Global Security: A Strategy Paper for the World Bank*. World Bank, Washington DC.

World Bank (2005) *Reducing Poverty on a Global Scale: Learning and Innovating for Development*, Washington.

World Bank (2007) *World Development Report 2008: Agriculture for Development* World Bank, Washington DC.

Xing, L., Fan, S.G., Luo, X.P. and Zhang, X.B. (2006) *Village Inequality in Western China: Implications for Development Strategy in Lagging Regions*. DSGD Discussion Papers 31, International Food Policy Research Institute.

Xinhua Net (2006) Vice Minister Targets Agricultural Value Adding, September 5, 2006.

Yan, D. (1999) Review of the Environmental Management Regulations in China's Cattle and Beef Industries. Master Dissertation Thesis, Department of Natural and Rural Systems Management, The University of Queensland.

Yan, X.J., Deng, R. and Sun, B.C. (2006) *Zhongguo Xuchanpin Shengchan Chengben Yu Shouyi Fenxi (Analysis of China Animal Husbandry Products Production Costs and Revenues)* China Agriculture Publishing House, Beijing.

Yang, J.S. (2005) *Economic Studies on the Production and Consumption of Safe Vegetables*. China Agriculture Publishing House, Beijing.

Yin, R. (1994) *Case Study Research: Design and Methods (2nd Ed.)*. Sage Publishing, Thousand Oaks.

Yu, J.B. and Ke, B.S. (1992) *Zhurou, Shucai Shichang Gaige Yanjiu (Research on the Reform of Pork and Vegetable Marketing)*. China Science and Technology Publishing House, Beijing.

Yu, T.F. (2001) Start the 10th Five Year Plan New Development Phase of the Inner Mongolia Livestock Sector. *Neimenggu Xumu Kexue (Inner Mongolia Livestock Science)* 1, 1–4.

Zhai, X.L. and Han, Y.J. (2008) Jirou Jiage Xingcheng Guocheng Ji Lirun Fenpei Qingkuang Diaocha Baogao (Investigation of Chicken Price Transmission and Benefit Distribution of Chicken Chain). *Nongye Zhanwang (Agricultural Outlook)* 4, 5, 17–20.

Zhang, C.G. and Longworth, J.W. (eds) (2000) *Zhongguo Rouniuye De Xiaoyi Tantao (Efficiency in China's Cattle and Beef Industry)*. China Agriculture Scientech Publishing House, Beijing.

Zhang, C.G. (ed) (2006a) *Zhuanxingzhong De Zhongguo Xumuye: Qushi Yu Zhengce Tiaozheng (China's Livestock Industry in Transition: Trend and Policy Adjustment)*. China Agriculture Press, Beijing.

Zhang, C.G., Xiong, C.K. and Li, S.P. (eds) (2000) *Zhongguo Rouniuye (the Cattle and Beef Industry in China)*. Zhongguo Nongye Keji Chubanshe (Agricultural Scientech Publishing House of China), Beijing.

Zhang, X.H. (1998) An Analysis of Household Pig Raising Situation and Economic Return in China. *Chinese Rural Survey* 1, 53–61.

Zhang, X.H. (2006b) Developing Smallholder Livestock Production for Creating Rural Employment and Increasing Farm Income. In *International Conference on Livestock Services*, Beijing, April 17–19.

Zhang, X.H. and Lu, M. (1997) Pig Production Scale and Feed Conversion Ratios. *Chinese Rural Economy* 10, 63–66.

Zhao, Y.T., Brown, C.G., Waldron, S.A., Longworth, J.W. and Li, P. (2005) *Zhonguo Fangzhi Qiye Jingji Guanli Juece Fenxi Fangfa (Building Economic Decision-Making Capabilities of Chinese Wool Textile Mills)*. ACIAR Technical Reports No. 61, Australian Centre for International Agricultural Research, Canberra.

Zhou, G.H., Liu, L., Xiu, X.L., Jian, H.M., Wang, L.Z., Sun, B.Z. and Tong, B.S. (2001) Productivity and Carcass Characteristics of Pure and Crossbred Chinese Yellow Cattle. *Meat Science* 58, 4, 359–362.

Zhou, Z.Y. (2003) Feed Versus Food: The Future Challenge and Balance for Farming. In *The Livestock Revolution: A Pathway From Poverty*, Brown and A.G. (Eds.), ATSE Crawford Fund, Canberra.

Zhou, Z.Y. and Tian, W.M. (2005) Developments in China's Grain Policies. In: Zhou, Z.Y. and Tian, W.M. (eds) *Foodgrain, Feedgrain and World Trade*. Ashgate, Aldershot.

Zhu, X.G. and Tian, W.M. (1989) The System of Decision Making for Agriculture in China. *The Australian Journal of Chinese Affairs* 21, 161–170.